Corneal Angiogenesis

Gordon K. Klintworth

Corneal Angiogenesis

A Comprehensive Critical Review

With 14 Figures

Springer-Verlag
New York Berlin Heidelberg London

Gordon K. Klintworth, M.D., Ph.D.
Duke University Medical Center
Durham, North Carolina 27710
USA

Library of Congress Cataloging-in-Publication Data
Klintworth, Gordon K.
   Corneal angiogenesis: a comprehensive critical review / Gordon K.
Klintworth, author.
      p.   cm.
   Includes bibliographical references and index.
   ISBN-13:978-1-4612-7787-3

      1. Cornea — Blood-vessels — Growth.   2. Cornea — Pathophysiology.
   I. Title.
   QP477.8.K58   1990
   612.8′41 — dc20   90-47929

Printed on acid-free paper.

Camera-ready copy provided by the author.

9 8 7 6 5 4 3 2 1

ISBN-13:978-1-4612-7787-3   e-ISBN-13: 978-1-4612-3076-2
DOI: 10.1007/978-1-4612-3076-2

This monograph is dedicated to three giants of ophthalmology who stimulated my interest in corneal angiogenesis and whom I had the great pleasure of knowing and learning from: Norman Ashton, David G. Cogan, and the late Isaac C. Michaelson.

PREFACE

Despite its relatively simple structure the cornea possesses many unique properties. These attributes include its crystal clarity and avascularity in the healthy state. This normally transparent structure has been the focal point of my research endeavors for more than two decades and this monograph summarizes current knowledge about angiogenesis within this tissue as well as information about the related issue of the cornea's normal avascularity. The text includes my thoughts on the subject and tries to provide a comprehensive overview of the topic based on studies by a large number of investigators who were either concerned with corneal neovascularization in particular or angiogenesis in general. Research by myself and numerous collaborators is cited in various appropriate places. This research involved invaluable collaborative efforts with Eric Alexander, Perry J. Blackshear, Peter C. Burger, David B. Chandler, Timmie J. Conrad, Joseph M. Corless, Mark Culton, Fusen H. Erkel, Carl H. Fromer, Herbert J. Glatt, Edward C. Halperin, William L. Haynes, Michael J. Helms, Tetsuo Hida, J. Stuart McCracken, Pam B. Morris, Alan D. Proia, Paul A. Raskauskas, William A. Russell, Mark W. Scroggs, Ray N. Sjaarda, Clayton F. Smith, Chanes Suvarnamani, Judy L. Swain and Michael T. Vu. Our research efforts would not have been possible without the generous research grant support from the National Eye Institute (Research Grants RO1-EY-00146 and P30-EY-05722) and a Louis B. Mayer Scholarship from Research to Prevent Blindness, Inc.

Finally, I wish to thank my wife Felicity and my children for the patience and self-sacrifices that they have had to endure while I utilized time that should have been devoted to them by pursuing research endeavors and writing.

Gordon K. Klintworth

CONTENTS

ABBREVIATIONS

The following abbreviations are used in the text:

ADP	=	adenosine diphosphate
aFGF	=	acidic fibroblast growth factor
ATP	=	adenosine triphosphate
bFGF	=	basic fibroblast growth factor
CAM	=	chorioallantoic membrane
cAMP	=	cyclic adenosine 3'-5' monophosphate
cDNA	=	complimentary deoxyribonucleic acid
CMP	=	cytidine monophosphate
Cu	=	copper
DG	=	diacylglycerol
DNA	=	deoxyribonucleic acid
ECL-1	=	proliferation-inducing endothelial cell lymphokine
EGF	=	epidermal growth factor
ELAM-1	=	endothelial-leukocyte adhesion molecule 1
ENDO-GF	=	platelet derived endothelial-growth factor
ESAF	=	endothelial cell stimulating angiogenesis factor
FGF	=	fibroblast growth factor
FGF-5	=	fibroblast growth factor 5
γIFN	=	gamma interferon
GM-CSF	=	granulocyte-macrophage colony stimulating factor
HAF	=	human angiogenic factor
HBGF	=	heparin binding growth factor
HBGF-1	=	acid fibroblast growth factor
HBGF-2	=	basic fibroblast growth factor
HBGF-3	=	heparin binding growth factor-3, int-2
HBGF-4	=	heparin binding growth factor-4, hst/KS3
HBGF-5	=	heparin binding growth factor-5,
HETEs	=	hydroxyeicosatetraenoic acids
I	=	inositol
ICAM-1	=	intercellular adhesion molecule 1
ICAM-2	=	intercellular adhesion molecule 2
IL-1	=	interleukin-1
IL-2	=	interleukin-2
IL-6	=	interleukin-6
IL-8	=	interleukin-8
IP_3	=	inositol 1,4,5-triphosphate
LTB_4	=	leukotreine B_4
MDGF	=	monocyte/macrophage derived growth factor (basic fibroblast growth factor)
MHC	=	major histocompatibility complex
mRNA	=	messenger ribonucleic acid
P	=	phosphate
PAF	=	platelet activating factor
PAI-1	=	plasminogen activator inhibitor type 1
PDBu	=	β-phorbol 12,13-dibutyrate
PD-ECGF	=	platelet derived endothelial-growth factor
PDGF	=	platelet derived growth factor
PDGF-A	=	platelet derived growth factor A chain
PDGF-B	=	platelet derived growth factor B chain

PG	=	prostaglandin
PGI $_2$	=	prostaglandin I$_2$
PGE$_1$	=	prostaglandin E$_1$
PGE$_2$	=	prostaglandin E$_2$
PI	=	phosphatidyl inositol
PIP	=	inositol diphosphate
PIP$_2$	=	inositol 1,4,5-triphosphate
PKC	=	protein kinase C
PPD	=	purified protein derivative of tuberculin
R1	=	fatty acid (usually saturated)
R2	=	fatty acid (usually unsaturated)
TGF-α	=	transforming growth factor alpha
TGF-β	=	transforming growth factor beta
TNF	=	tumor necrosis factor
TNFα	=	tumor necrosis factor alpha
TNFβ	=	tumor necrosis factor beta
TPA	=	12-0-tetradecanoyl phorbol-13-acetate
VCAM-1	=	vascular cell adhesion molecule 1
VEGF	=	vascular endothelial growth factor
VPF	=	vascular permeability factor

Chapter 1

CORNEAL AVASCULARITY AND VASCULARITY

The cornea normally lacks blood vessels, but in numerous diverse natural and experimental situations, capillaries extend into this tissue from the pericorneal vascular plexus. The avascularity of the normal cornea and its vascularization in certain pathologic states have attracted attention for at least the better part of almost two centuries (372,767). During the early part of the twentieth century the cornea was used in studies of angiogenesis in humans (28,102,428,430) and during the 1920's and 1930's animals were used to study new vessel formation in the cornea (191,389-391,733). However, the first attempts to investigate the pathogenesis of corneal vascularization in depth were only made in 1949 by Campbell and Michaelson (116) and Cogan (143).

Corneal neovascularization usually extends between the collagen lamellae into the substantia propria of the cornea, but it is also a component of a fibrovascular sheet between the corneal epithelium and Bowman's layer (a pannus). An understanding of the mechanism of this common clinical event is of fundamental biological importance and relevant to many physiologic and pathologic conditions, including such apparently diverse entities as diabetic retinopathy, iris neovascularization, the retinopathy of prematurity, tumor growth, wound healing, and luteinization of the ovary. The failure of some transplanted tissues to elicit an adequate angiogenic response may play a cardinal role in the inability of fragments of tissue such as liver and kidney to survive when implanted into ectopic sites (745). Conversely, the induction of blood vessels may explain why bone marrow can survive after transplantation to ectopic sites (744).

Vascularized corneas are clinically significant because they are associated with a diminished visual acuity, an increased risk for graft rejection (345,448) and are prone to opacification from the deposition of lipid (144). The presence of blood vessels within the cornea also influences other aspects of this unique tissue. For example, during the healing of a total corneal epithelial defect from the epithelium of the surrounding conjunctiva the migrating epithelium normally becomes transformed into a cornea-like epithelium with loss of goblet cells. However, if the cornea is vascularized during, or after, this healing period the conjunctival transdifferentiation does not take place and goblet cells persist in the epithelium overlying the cornea (262,471,508,751,768,769) and the barrier function of the epithelium remains impaired (365). Despite being associated with corneal clouding and diminished visual acuity, the mere existence of blood vessels within the cornea is probably not the cause of the opacification, because certain vascularized tissues, such as the reptile spectacle, remain transparent even in the presence of blood vessels (526). Also, corneas with established experimental angiogenesis are often transparent.

Aside from these adverse effects the presence of blood vessels appears desirable in situations where the integrity of the anterior segment needs to be preserved as after corneal burns (101,360). Indeed as long ago as 1929 the induction of corneal angiogenesis by cauterizing a line from the corneoscleral limbus was advocated in the treatment of indolent ulcers (656). This view is supported by the experimental observation that stromal ulceration occurs less readily in vascularized corneas than in those that are avascular following thermal burns (153).

While normally avascular, except at its extreme periphery, throughout most of the animal kingdom the cornea is alleged to be vascularized nonpathologically in the Rocky mountain bighorned sheep (*Ovis canadensi Shaw*), armadillo, manatee, and in some Japanese salamanders, such as the giant 5 foot long salamander *Megalobatrachus maximus* (419,440,510,733,746). The peripheral cornea of some fishes (Amia, Cobitis, Gobius and Mola) is vascularized (569). Blood vessels appear to invade the cornea spontaneously in gray Norwegian rats, but not in Wistar albino rats, and the incidence of this vascularization increases with age for reasons that remain unexplained (460). The presence of blood vessels within apparently normal corneas has also been noted in athymic, nude mice (nu/nu) and the euthymic, hairless mutant mouse strain (SKH1; hr/hr) (338,571). The aberrant corneal vascularization in these mice is not related to immunodeficiency because the corneas from the severe combined immunodeficiency mutant mouse strain are avascular and the euthymic hairless mutant mouse strain (SKH1; hr/hr) is not immunodeficient despite the presence of corneal blood vessels (571). However, why nude mice vascularize remains unclear, because the fenestrated and non-fenestrated subepithelial capillaries in the central cornea of these mice is associated with occasional polymorphonuclear leukocytes (338), suggesting that the vascularization may be associated with inflammation as discussed elsewhere. Moreover, several years ago we failed to find blood vessels within the cornea of athymic nude mice, but discerned that they were prone to enhanced corneal angiogenesis following cauterization with silver/potassium nitrate (703). However, the normality of the corneal tissue in all of the aforementioned animals has not been adequately documented. The gecko is stated to have a vascularized cornea (510), but the vascularized preocular tissue in this lizard is not the cornea. It is the reptile spectacle - a structure located external to the cornea (526). Until blood vessels are demonstrated within the cornea of healthy normal animals with no antecedent corneal trauma, infection, or other inciting causes of angiogenesis claims of non-pathologic corneal neovascularization can only be viewed with skepticism. Nevertheless, it is important that the issue be resolved, because if true it could shed light on our understanding of why the normal cornea is avascular and the related question of why this tissue vascularizes under certain pathologic conditions.

Blood vessels enter the human cornea at variable depths and produce a variety of vascular patterns. The depth of the vasculature within the cornea relates to the inciting pathologic process (184). Clinically the vascularization may extend from the conjunctival superficial limbal plexus into the subepithelial portion of the cornea without actually invading the corneal stroma. Such superficial vessels may extend from the entire limbus like a cloth (pannus) or the vessels may form a small focal bundle (fasciculus). When a dense mass of vessels grows towards the cornea the epithelium becomes elevated in a gelatinous pink mass (epaulet). The appearance of pannus varies with the cause and ophthalmologists have traditionally recognized several types: pannus trachomatosus, pannus leprosus, pannus phlyctenulosus, pannus degenerativus, and glaucomatous pannus (184). An invasion of the human corneal stroma by blood vessels derived from the anterior ciliary arteries (interstitial vascularization) also results in several clinically distinct types of vascular patterns: terminal loops, brush-like vessels that have been likened to branchings of birch twigs, an umbel form in which the central vasculature assumes stellate branches, a lattice form with right angled budding, interstitial arcades derived from episcleral vessels and aberrant vessels that traverse the corneal stroma irregularly (184).

Different patterns of angiogenesis also form under experimental conditions. For instance two days after intracorneal grafts of semi-allogeneic lymph node fragments into mouse corneas Muthukkaruppan and Auerbach (563) observed prominent vasodilation in the limbal region immediately adjacent to grafts and this was followed by an extension of vascular loops into the cornea towards the lymph node fragments which did not become fully vascularized. In contrast tumors (S180 sarcomas and C755 mammary tumors) induced a brush-like fronse of capillary sprouts from the apices of capillary loops and these extended into the tumors.

Chapter 2

CAUSES OF CORNEAL NEOVASCULARIZATION

A wide variety of disorders of the cornea culminate in angiogenesis and a considerable amount of information has accumulated about the circumstances under which newly formed blood vessels sprout and extend centripetally into the cornea. In man corneal neovascularization complicates numerous infections (including trachoma, herpes simplex keratitis, leishmaniasis and onchocerciasis), immunologic processes (Stevens-Johnson's syndrome, graft rejection), alkali burns, as well as traumatic, inflammatory, toxic and nutritional deficiency states and the wearing of some contact lenses (11,104,130,184,190,273,388,638,645). It is noteworthy that corneal vascularization is relatively uncommon in fungal keratitis (184).

Under experimental conditions blood vessels can be made to invade the cornea in a wide variety of ways, including chemical, microbiological, and physical injuries, as well as toxic and nutritional deficiency states, and immunological reactions (Table 1). A notable feature of the different experimental models of corneal vascularization is the variability in the incidence and severity of the induced corneal vascularization. In some experimental models, such as corneal cauterization with silver/potassium nitrate or the intracorneal instillation of interleukin 1, corneal angiogenesis occurs with high frequency. In other situations the incidence is much lower. For example, in zinc deficient rats Follis, Day and McCollum (248) observed corneal vascularization in only 2 of 7 rats (1 killed at 6 weeks, the other at 11 weeks). In some experimental situations, such as ascorbic acid deficiency (730), complete starvation (65), vitamin A deficiency (263), or the intracorneal installation of saline (419,822), blood vessels do not constantly invade the cornea. Particularly potent stimulators of corneal neovascularization are immunologically provoked reactions such as the intracorneal instillation of an antigen into sensitized animals (263) and graft rejection (345).

It is noteworthy that when corneal angiogenesis is induced experimentally by nutritional deficiencies, the neovascularization may not be a direct consequence of the deficiency state but a complication of it. The corneal neovascularization that is readily induced in rats by vitamin A deficiency, can be prevented by housing the vitamin A deficient rats in a germ-free environment and feeding a diet supplemented with low levels of retinoic acid, the fully oxidized form of Vitamin A (120).

Chapter 3

SEQUENCE OF EVENTS IN CORNEAL NEOVASCULARIZATION

Numerous investigators have studied the ultrastructure of growing blood vessels has been studied in the cornea of the rat (377,666,667,736,737) and rabbit (504,505,728,817,818) but the chain of events that precedes and follows the induction of corneal neovascularization has been studied in relatively few experimental models (18,63,106,108,116,143,194,263-265,416,524,666,694). The endothelial cell-specific tubular organelles (Weibel-Palade bodies) that contain a specific glycoprotein involved in blood coagulation (von Willebrand factor) (523) have been observed in growing corneal blood vessels (326,504).

The mechanisms involved in corneal vascularization remain ill defined, but as in angiogenesis elsewhere the process by which blood vessels grow into the cornea can be divided into several discrete but overlapping phases.

1. Latent Period between Corneal Injury and the Onset of Angiogenesis

Attempts have been made to unravel the complex chain of events that take place during its early phases of corneal neovascularization (194,417,524,751). The evolution of the early events in corneal vascularization has been most extensively studied by light microscopy, transmission electron microscopy, autoradiography, scanning electron microscopy of methylmethacrylate casts, and fluorescein angiography of the pericorneal and corneal blood vessels in the rat following a focal chemical cautery (silver/potassium nitrate) (Table 2) (106-108,263,264,524,782). In this model the pericorneal blood vessels dilate markedly within 1 hour of corneal cauterization and during the subsequent 5 hours leukocytes and platelets become conspicuous within the dilated and permeable vascular channels. Polymorphonuclear leukocytes reach the extravascular space from the microvasculature within the next 24 hours and migrate through the corneal stroma towards the site of cauterization. Together with the accompanying corneal edema these inflammatory cells opacify the cornea markedly. In other experimental models of corneal vascularization similar events take place, but the nature of the cellular infiltrate varies and so does the length of time that precedes the onset of neovascularization.

The first new vessels appear as sprouts from the capillary arcade and postcapillary venules 27 hours after corneal cauterization.

The sequence of events in the preangiogenic period following corneal injury has been strikingly similar in different experimental settings despite differences in the models, but the latent period between the initial angiogenic provoking manipulations and the beginning of neovascularization varies considerably in different models. In some the vascular endothelial cells begin to migrate and proliferate within 24 hours, while in other situations the onset of angiogenesis takes days or weeks. The interval between the initiating event and the onset of the neovascularization relates directly with the timing of the initial intracorneal leukocytic infiltration (263). Situations in which the leukocytic invasion is delayed (such as Vitamin A deficiency) have a late onset of corneal neovascularization, whereas injuries that provoke a rapid and extensive leukocytic invasion elicit an early and enhanced corneal neovascularization.

In several experimental situations in which corneal vascularization can be induced, Alessandri, Raju and Gullino (9) detected elevated sialic acid levels in the

5

cornea prior to the time that the tissue would be penetrated by capillaries, and they speculated that this might reflect an increase in the corneal ganglioside content. However, sialic acid, which is derived from various sources, is a component of tears (434) and could have been the source of this terminal constituent on carbohydrate side-chains of numerous glycoproteins.

2.   Vascular Dilatation, Increased Vascular Permeability and Corneal Edema

Stromal edema not only accompanies corneal neovascularization, but precedes and apparently facilitates the growth of blood vessels into the cornea (143,182,512). The stromal edema that antecedes corneal neovascularization appears to be of the inflammatory type caused by an increased permeability of the pericorneal vasculature, rather than edema due to corneal endothelial cell dysfunction or destruction. The increased vascular permeability associated with the inflammatory response occurs primarily from the postcapillary venules and this occurs between venular endothelial cell junctions probably because of endothelial cell contraction. A wide variety of inflammatory mediators contribute to this increased vascular permeability. The compounds include histamine, 5-hydroxytryptamine (serotonin), bradykinin, plasma kallikrein, substance P, adenosine diphosphate (ADP), adenosine, inosine, prostaglandins (E_1, E_2, $F_{2\alpha}$), leukotrienes (C_4, D_4, E_4, B_4), components of complement cascade (C3a, C5a), platelet-activating factor-acether, fibrin-derived peptides, free radicals, ischemia, and immune aggregates. The effect can be counteracted by several autacoids (such as epinephrine, vasopressin and cortisone) and a variety of drugs (such as β_2 receptor agonists, antihistamine, glucocorticoids and xanthines) (732). Topical prostaglandin E_1 and the histamine liberator 48/80 increase the permeability of pericorneal blood vessels (in the rat), but this effect has not been found after topical bradykinin and histamine diphosphate (579).

The edema continues during the angiogenic process while the invading new vessels remain permeable to fluid, plasma proteins, and smaller molecules (107,579,736) and this persists until the newly formed vascular endothelial cells become united and fully develop a basal lamina with supporting cells.

3.   Endothelial Cell Activation with Retraction and Decreased Endothelial Cell Junctions

The pericorneal vessels are normally impermeable to carbon but do leak thorium dioxide and fluorescein, at least in the rat (579,735,736). Developing corneal vessels lack firm intercellular junctional complexes, and are particularly permeable to fluorescein, carbon, brilliant benzo blue 6A (Chicago blue), and thorium dioxide (107,143,579,736). In contrast, established vessels within corneas in advanced stages of wound healing lack this leakiness and thorium dioxide does not escape from them (736,737). Transmission microscopy has disclosed the leakage from newly formed corneal blood vessels to be between endothelial cells (667). In common with newly formed blood vessels elsewhere, those within the cornea are abnormally permeable (666).

While the initiator of corneal angiogenesis remains unknown the endothelium of the pericorneal vasculature responds soon after corneal injury as evidenced by DNA synthesis within these cells. This begins within 21 hours of corneal cauterization (108,524). By 24 hours after injury, the pericorneal vascular endothelial cells and pericytes manifest prominent nucleoli and abundant polyribosomes. The initial ultrastructural alteration with the endothelium of pericorneal blood vessels following chemical cauterization of the cornea is cellular enlargement, nucleolar prominence,

6

and dispersion of polyribosomes into their free form (32,524). Following an angiogenic provoking injury the pericorneal vascular endothelial cells become plumper and some of the cytoplasmic organelles become more prominent and intercellular junctional complexes disappear. The initial direct evidence of new vessel formation appears as enlargement of the nuclei and nucleoli of the vascular endothelium and the incorporation of H^3-thymidine within the nuclei of these same cells.

4. Degradation of Endothelial Cell Basal Lamina

At the onset of new vessel formation the basal lamina surrounding the vascular endothelium becomes degraded allowing the endothelial cells to migrate into the extracellular matrix. This destruction of the basal lamina within the wall of the responding blood vessels is presumably due to the action of proteolytic enzymes (including plasminogen activator, and type IV collagenase) that are derived from one or more potential sources (the vascular endothelium, polymorphonuclear leukocytes, mononuclear phagocytes, mast cells, corneal epithelium, microorganisms). The growing vascular endothelium itself may contribute to its ability to invade tissues, since vascular endothelial cells produce activators of plasminogen and other proteases in culture (479,630,658). Cultured bovine endothelial cells from capillaries, but not the aorta, synthesize plasminogen activator and latent collagenase when stimulated with substances having an angiogenic potential (bovine retinal extract, mouse adipocyte conditioned medium, human hepatoma lysate and 12-O-tetradecanoyl phorbol-13-acetate) (TPA). A collagenous substrate seems to be important for capillary endothelial growth and Shor and colleagues (670,671) have suggested that this relates to the capacity for binding the angiogenic factor.

5. Endothelial Cell Migration and Replication

The assembly of new vessels depends upon the independent phenomena of endothelial cell migration and endothelial cell replication, which seem to take place simultaneously. In studies of corneal angiogenesis in the rabbit, Yamagami (817,818) observed that cytoplasmic processes of the endothelial cells at the tip of the growing new vessels extended toward the injured area, while mitoses were restricted to the more proximal cells on the capillary buds (817,818) and postulated that vascular endothelial cells migrate and form provisional lumens before they undergo cell division. Observations on corneal neovascularization in rats are consistent with this view, but because of the difficulty in timing initial mitoses in tissue sections it has not been possible to determine whether endothelial cell migration precedes or succeeds endothelial cell proliferation. Mitotic figures have been noted in pericorneal vascular endothelial cells 36 hours after injury, whereas the first buds are evident by scanning electron microscopy of methylmethacrylate casts of newly forming blood vessels 9 hours earlier following cautery (106). However, the initial ^{3}H-thymidine label in vascular endothelial cells (in the S phase) occurs by 21 hours (108). Allowing a minimum of 2 hours for completion of the S phase of cell division, and an additional 2 hours for the G_2 phase, the first mitotic cell divisions could be completed at approximately 26 hours after cautery or at about the time of the initial vascular buds.

Under normal circumstances the pericorneal vascular plexus, like the endothelium of other blood vessels, rarely undergoes cell division or incorporates H3-thymidine (108). This is because cells of the vascular endothelium are believed to have a normal turnover time of 2 months or longer due to a prolonged G_1 phase or time spent out of the cell cycle (G_0 phase) or both (46). Most pericorneal vascular

7

endothelial cells presumably become triggered to undergo mitosis while they are in the G_1 or G_0 stage of the cell cycle. With certain corneal injuries, such as chemical cautery, the mitogen reaches the vascular endothelium soon after the injury, as evidenced by morphologic changes and the incorporation of H^3-thymidine, an indication of entry into the S phase of the cell cycle, by the vascular endothelium within 24 hours of corneal injury (108,524). Because the combined duration of the S and G_2 phases of the cell cycle in most cell types lasts about 10 to 12 hours, one would anticipate the first mitotic figures within endothelial cells within 36 hours of injury in this model, and indeed this occurs.

Vascular endothelium in common with other cells loses its capacity for mitosis after appropriate doses of x-irradiation, but yet retains the ability to migrate in response to a chemotactic stimulus. This property has been shown in cell culture systems (693,789), and also appears to operate *in vivo*. Following ocular irradiation, chemical cautery to the cornea still evokes endothelial cell migration and vascular sprouting, albeit less than normal, despite the absence of cell division (290,695).

Although the role of pericytes in angiogenesis remains uncertain (32,126,524,666), these cells, which may also contribute to the formation of new vessels, manifest prominent nucleoli and abundant polyribosomes and become labeled with H^3-thymidine during corneal neovascularization elicited by chemical cautery (108,524). They also incorporate thymidine during angiogenesis in the skin following thermal injury (689) or implantation of tumor cells (125,126). The origin of pericytes in corneal angiogenesis remains uncertain and corneal fibroblasts (keratocytes) have been implicated (565).

6.   Vascular Sprouting

Clearly the endothelial cells in different components of the microvasculature do not contribute equally to the angiogenic response. As shown by light and electron microscopic studies of early angiogenesis in the rabbit ear chamber (142) and the cornea (32), the newly formed vascular sprouts that give rise to corneal blood vessels stem largely from venules and preexisting capillaries in the pericorneal vascular plexus and not from small arteries or arterioles. Investigations by both autoradiography and scanning electron microscopy of methylmethacrylate casts of newly formed corneal blood vessels have illustrated this point dramatically (106,108). Also, the endothelium of small arteries is also not conspicuously labeled with H^3-thymidine during neovascularization of the skin (689). The reason why new vessels should arise predominantly from only a certain part of the microvasculature remains to be determined, but it is noteworthy that the emigration of leukocytes also occurs predominantly in postcapillary venules (559). Since the interstitial fluid normally drains into the venous side of the microcirculation it may reflect a tendency of an angiogenic substance(s) to drain with the extracellular fluid.

7.   Vascular lumen formation and anastomoses

The new blood vessels, which are evident in histological sections as early as 33 hours after injury, emerge from the pericorneal plexus and extend toward the site of the corneal injury, which they reach within 72 hours of the initiating stimulus. A maze of proliferating channels set in a sea of inflammatory cells extend into the cornea from the pericorneal vasculature. After the formation of blind vascular channels the newly formed capillaries become linked to each other establishing a circulation.

8.   Formation of Basal Lamina Around Newly Formed Vessels

In contrast to established blood vessels, which have a continuous basal lamina and one or more layers of pericytes (737), the endothelium of forming new vessels lack this supporting scaffold until they secrete components of the extracellular matrix in a temporally ordered sequence (251). In an immunoperoxidase study of corneal vascularization induced in the mouse by silver nitrate cautery Form, Pratt and Madri (251) noted laminin throughout the newly formed vessels, as well as in individual cells at the migrating, proliferating tips. In contrast type IV collagen correlated with lumen formation and was not detected at the vessel tips.

9.   Phase of Capillary Regression and Vascular Maturation

The eventual fate of blood vessels once they enter the cornea remains ill defined, but a persistent stimulus is required for their maintenance. For example, following the implantation into the rabbit cornea of a pellet containing tumor cells, blood vessels grow into the cornea, but they regress within two weeks of its removal. In the rat cornea, within a week following silver/potassium nitrate cauterization, the vascular lattice simplifies as many newly formed channels apparently resorb, while other vessels enlarge and extend as loops to and from the site of injury. Smaller corneal vessels begin to disappear in favor of fewer larger vascular channels at 7 days in this model. Stereoscopic views of scanning electron micrographs of methacrylate casts of the induced corneal microvasculature 9 days after injury has disclosed that the large intracorneal vessels are continuous with either a pericorneal artery or vein (106). The large vessels possess flat endothelial cells and lack smooth muscle (venules), whereas the arterioles develop thicker walls and are lined by endothelial cells as well as mural cells with microfilaments focally arranged with irregular densities (106). By 1 month following the resolution of the corneal edema and inflammatory infiltrate the cornea regains much of its original clarity, except for the opaqueness due to the vascularity. With the aid of scanning electron microscopy of vascular casts, the surviving vessels receive blood directly from the pericorneal arterioles. Fluorescein angiography provides the means whereby the direction of blood can be determined with ease. With this technique the surviving vessels develop morphologic attributes of arterioles, whereas vessels draining into the pericorneal venules acquire features of venules (107). The vasculature about a healed corneal lesion may appear inconspicuous, but become apparent again when vasodilatation is induced, as with acute inflammation provoked by an additional new corneal injury. The site of some vessels persist as non-blood filled channels ("ghost vessels"). In the chemically cauterized rat cornea some of the large vessels passing between the corneoscleral limbus and the site of injury may persist for as long as 6 months (107).

Corneal angiogenesis is seldom, if ever, an isolated phenomenon. Aside from vascular endothelial activation other cells react during corneal angiogenesis. Ultrastructural and autoradiographic studies during the early stages of corneal vascularization indicate that the vascular endothelium and pericytes of pericorneal blood vessels synthesize DNA and undergo mitosis at about the same time as the endothelium, epithelium, and fibroblasts of the cornea (108,524). When blood vessels invade the avascular cornea in different experimental models other cell types that do not normally reside in this tissue participate in the event.   Leukocytes (polymorphonuclear leukocytes, monocytes, lymphocytes) and mast cells invade the cornea (707,709,831). Melanocytes (343,525) also frequently establish residency within the cornea. Some Langerhans' cells reside normally in the peripheral cornea, but they increase in density and involve the central cornea in inflammation (280,466). Stimuli that provoke corneal neovascularization also elicit Langerhans cell migration

9

(301,646,739,807), but the two phenomena are not interdependent (571). In addition, lymphatic vessels invade the cornea following external corneal injury (148) or intracameral alloxan (147). These channels, whose ultrastructure has been described in the rabbit at different stages of development (150,151), provide a route whereby albumin and other large molecules can gain access to the lymph nodes in neck (149). Another cell type that has been observed in association with corneal blood vessels is the Schwann cell and its accompanying neurites (462).

With certain non-progressive corneal injuries the regression of intracorneal blood vessels can be accelerated or promoted by argon laser photocoagulation (87,133,134,621), as well as by the induction of thrombosis by the intravenous administration of photosensitizing dyes, such as rose bengal (156,366,367,529,530) or dihematoporphyrin ether (201,830) followed by argon laser irradiation. Therapy with argon photocoagulation, which has been found to be effective apparently in non-controlled isolated human cases (763), poses clinical problems because the thermal effects may worsen the neovascularization (573) or result in corneal thinning and damage to the underlying iris (499). In an inadequately controlled clinical study cryotherapy is reputed to accelerate the regression of corneal vascularization (335).

Chapter 4

QUANTITATION OF CORNEAL VASCULARIZATION

Numerous methods have been used to quantitate corneal angiogenesis but almost all lack sufficient precision to provide reliable data suitable for biostatistical analysis. Many investigators of corneal neovascularization have only indicated whether a particular injury or intracorneal inoculation did or did not induce angiogenesis. In some studies the incidence of an angiogenic response has been used as an indication of the angiogenicity. Because it has become abundantly clear that the frequency and degree of corneal neovascularization varies in different situations the extent of induced corneal vascularization must be quantitated for meaningful comparisons between different experimental groups and their appropriate controls. To achieve this end some authors have graded the degree of corneal vascularization subjectively (61,174,266). Such methods have included the measuring of the vascular ingrowth from the corneoscleral limbus with a caliper and grading the extent of the vascularization subjectively (266); others have based the grading system on the number and length of the vascular loops (174) or on the distance that the blood vessels extended towards a burn from the corneoscleral limbus (494). Surprisingly, many authors have either not described their grading system or have simply illustrated representative photographs. A commonly used non-objective scoring of corneal vascularization on a 1+ to 4+ basis is insensitive and difficult to standardize. Efforts have also been made to grade angiogenesis according to the density of vascular growth (385). Attempts have also been made to rank conventional corneal photographs, but this method also suffers from several inherent weaknesses, which include the difficulty in obtaining adequate limbus-to-limbus photographs of the entire cornea with satisfactorily resolved views of the blood vessels. Moreover, this technique is not applicable to the small eyes of rodents with current technology. Some investigators have attempted to measure the mean or maximum length of the new vessels within the cornea (218,354,401,402). How this is achieved is frequently not stated, but some authors have attempted to measure vessel length in certain positions, such as perpendicular to the angiogenic lesion. However, because of the corneal curvature such measurements probably reflect the longest radial length of the cornea with neovascularization rather than actual vessel length because vessel growth is linear and perpendicular to the limbus. The area of corneal neovascularization, as a percentage of total corneal area, has been estimated by performing planimetry on projected photographs of corneas (776). Many researchers have recognized the shortcomings of vessel length as a measure of corneal neovascularization. For example the linear growth rate in some experimental situations may not differ substantially from controls despite the induction of a denser and broader growth of vessels (218). Attempts have been made to take the extent of the angiogenic response into account. This has been tried by adding the total length of all the new vessels (155), but this still underestimates the degree of neovascularization. Another approach to compare the angiogenic response in different experimental groups is to determine the time taken for blood vessels to reach a particular lesion (32). Aside from the difficulty in determining the length of vessels that grow in by a circuitous route, measurements limited to vessel length totally ignore the considerable variability in the number of new vessels in different situations. A measure of the area of corneal neovascularization is clearly more desirable. This has been attempted by measuring the vessel-producing limbal base and the length of the leading vessel in india ink perfused post-mortem corneas and calculating the obtained triangular area (535). Similar areas have been calculated in living rabbits (55). These determinations of vascularized areas suffer from

11

significant weaknesses. Firstly, the area of interest is not completely filled with blood vessels. Secondly, the areas are calculated from linear measurements with the same inaccuracies implicit in those measurements. Indeed this has been estimated by measuring the length of the vessels and the width of the limbal base with calipers and multiplying these two dimensions (204). Although such calculations provide numbers they lack precision and are not true representations of areas. In fatal animal experiments some investigators have enhanced the visibility of the induced corneal blood vessels by perfusing the circulation prior to, or immediately after, death with colloidal carbon and then measured the vascularized area of the mounted specimens of the cornea with methods such as a microscope eye piece containing a micrometer with a Vernier moving stage (116), by stereologic means (291), computerized planimetry (291), or computerized image analysis (612). Different investigators have approached the problem of how to quantitate the carbon-filled vessels. For example, in an early study, Sholley, Gimbrone and Cotran (694) measured the length of the vessels within flat preparations of rat corneas with a calibrated dissecting microscope at eight points equally spaced around the limbus and averaged these measurements to produce a single figure for the vascular length of each cornea. In subsequent studies Sholley and colleagues (692) used photographic negatives of colloidal carbon-perfused whole-mount preparations to measure the length of the vascular penetration into the cornea of rats. For quantitation of ingrowth at 4 days after corneal injury, 35 mm negatives were projected onto the screen of a microfische reader at a known magnification. For quantitation of the ingrowth at 7 days, 4 x 5-inch magnified negatives were transilluminated and measured directly. Two random, radially orientated measurements were taken, using a Vernier caliper, from each corneal quadrant, starting at the most peripheral vascular cascade and ending at the tip of the longest vessel along the radius selected. The eight measurements were averaged to provide a single length measurement for each cornea.

Because reliable methods of quantitating the variable amounts of corneal vascularization under different circumstances are critical for studies that evaluate factors that influence new vessel growth in this tissue, we are actively seeking better methods to measure corneal angiogenesis in the living animal. To date the most reliable and objective method for quantitating corneal angiogenesis involves computerized image analysis of corneal flat preparations that are made postmortem after vascular perfusion with colloidal carbon (162,337,612,647,731).

Chapter 5

USE OF THE CORNEA TO ASSAY FOR ANGIOGENIC ACTIVITY

A major difficulty facing investigators of angiogenesis has been the lack of a specific, sensitive, in vivo bioassay system for angiogenic factors. A wide variety of systems have been used in assays for angiogenic factors. These include implantations into rabbit ear chambers (138,659,815), hamster cheek pouch chambers (416,674), the chick chorioallantoic membrane (CAM) (23,33,44,223,241,482,642,686,757), the rat kidney capsule (336,623), polyester sponges (10), healing rabbit bone grafts (154,200), the subcutaneous air "pouch" of Selye (625,626,643), epididymal fat pads seeded onto confluent cultures of myofibroblastic cells (660) as well as the subcutaneous implantation of polyvinyl alcohol sponge discs (211,821) or alginate beads containing sequestered cells (602). Vascular endothelial cell cultures derived from different sources have also been used to assess cell replication, DNA synthesis, cell migration, capillary-like tubular structure formation (239,821) as well as other cellular responses to putative angiogenic and anti-angiogenic factors. All of these methods suffer from significant shortcomings not the least of which is a difficulty in quantitation. Assay systems that involve cultured vascular endothelial cells also pose major problems. Because endothelial cells represent a heterogeneous population (210) the responses by certain endothelial cells to biologically active substances may not necessarily reflect reactions of the microvasculature from which new blood vessels arise.

Because of its relatively large size, ease of accessibility, and normal avascular state, the rabbit cornea is a popular tissue in which to assay angiogenic and anti-angiogenic activity. The misconception of the cornea being an immune privileged site enhanced this popularity. The angiogenic response to various substances has been determined following their injection or implantation into the transparent cornea (Table 2). Some substances of interest have been injected directly into the cornea through a thin needle (8,9,18,55,63,92,128,143,191,195,209,224,258,297,325,374,468, 513,515,522,532,551,617,742,829). Portions of tumor and various tissues have commonly been introduced into the cornea after making a small incision on the corneal dome to about one half of the corneal thickness and producing a pocket or tunnel for the specimen with a delicate spatula (58,215,240,283,353,385, 443,523,617,827). To obtain a effective angiogenic response in such assays the bottom of the corneal pocket needs to be within a critical distance of the corneoscleral limbus (2±0.5 mm). Other assays have employed the intracorneal instillation of substances incorporated into inert "sustained release" polymers such as those made from ethylene vinyl acetate (Elvax 40) (DuPont) (55,520,522,827), polyhydroxymethacrylate (Hydron) (Hydron Labs) (443), or methylcellulose (7,223,475). A continuous concentration gradient of the test substance can also be achieved by the more tedious intracorneal infusion via a fine intracorneal polyethylene tube connected to a slow release Alzet minipump implanted beneath the scalp (197). Disadvantages of the rabbit cornea include the expense and time consuming nature of the assay, problems in quantitation and difficulties in interpretation caused by the lack of a genetically homogenous recipient population. Consequently, the cornea of the rat (154,252,265,413,456,609,610), mouse (204,563,564) and guinea pig (607,608) have also been used to a lesser extent in comparable assays. Inoculations into inbred animals, like the mouse, allow corneal neovascularization to be studied under genetically controlled conditions. By controlling for the sex, age and genetic background of the recipient the influence of certain variables can be avoided. Moreover, the vast body of information about both the murine genome and its mutations together with knowledge about immunologic

13

mechanisms in the mouse offers a distinct advantage over the rabbit, aside from the significantly lower cost of each analysis. However, the disadvantages of the murine cornea include the greater difficulty of introducing substances into it and of making clinical observations and photographs of the angiogenic response.

Despite the attractiveness of the cornea as a bioassay for angiogenesis caution is needed in the interpretation of the results. Moreover, animals may also scratch at the cornea and traumatize the site of innoculation especially if it is irritative or painful. When corneal neovascularization has followed an inoculum angiogenic activity has commonly been attributed to it, despite the fact that such a response does not necessarily indicate that the inoculum possesses direct angiogenic activity. As can be seen from Table 2 many viable and nonviable foreign materials provoke angiogenesis when introduced into the cornea. Many of the inoculations have involved antigenic material that is foreign to the host and an inducer of an immunologic and inflammatory response. The vast majority of substances that evoke angiogenesis when introduced into the cornea clearly do not solely and specifically provoke new vessel formation. In my experience the intracorneal instillation of almost any foreign substance elicits a nonspecific, but variable, inflammatory reaction, and indeed tissue surrounding an inoculum is usually infiltrated with leukocytes regardless of the nature of the introduced material.

The study of Ryu and Albert (654) illustrates the problem of trying to infer angiogenic activity to something implanted into the cornea. They found that the implantation of both viable and nonviable tumor cells (melanoma and retinoblastoma) into the corneal stroma produced a nonspecific localized keratitis in which the extent of corneal neovascularization correlated with the degree of inflammation. Importantly, a significant difference between the corneas containing live or dead tumors was not observed. Moreover, in rabbits made immunodeficient by radiation, the implanted tumors induced a negligible inflammatory cell infiltrate and negligible vascularization. Ryu and Albert (654) found that the vascular invasion of the cornea following tumor implantation was not modified by denaturation of the transplanted tumor by boiling or formalin fixation, both of which might be expected to inactivate a tumor angiogenic factor or to prevent its synthesis.

An additional difficulty facing investigations of potential angiogenic factors is the fact that positive responses are not always consistent in different assay systems. For instance some substances, such as formyl methionyl leucyl phenylalanine, that induce corneal angiogenesis after they are introduced into the cornea fail to be angiogenic in the chick chorioallantoic membrane assay and do not induce aortal or capillary endothelial cell migration (522). Observations like this stress the importance of not relying solely on intracorneal instillations for conclusions about angiogenic activity.

Substances Not Provoking Angiogenesis after intracorneal instillation

Despite the frequency with which neovascularization is induced by the intracorneal instillation of a wide variety of substances, everything that is introduced into the cornea does not provoke angiogenesis (Tables 3 and 4). Some substances, such as neonatal scapular cartilage, that have not evoked an angiogenic response have elicited minimal or no apparent inflammatory response in the cornea (93). Scant attention has been given to observations of this type, yet an understanding of why some substances do not elicit an angiogenic response, or do so inconsistently is critical to our understanding of why a myriad of other intracorneal instillations evoke angiogenesis. In some instances a negative angiogenic response may reflect leakage of the inoculum from the cornea, an insufficient time of observation, or an inoculum beyond the critical distance from the corneoscleral limbus. A variable response with some substances may be due in part to genetic heterogeneity in the host or to difficulties inherent in the technique. It is extremely difficult, if not impossible, to consistently introduce the same number of cells or an identical quantity of a substance into the same part of the cornea due to leakage from the point of instillation. Aside from this there are the inherent problems of quantitating corneal vascularization in the cornea. Also, the dose of the inoculation is sometimes critical to whether an angiogenic response will, or will not, occur (55,419). BenEzra (55), for example, elicited an angiogenic response after the intracorneal implantion of Elvax pellets containing 10 µg of either fibroblast growth factor or epidermal growth factor, but did not observe corneal neovascularization when the implants only contained 1µg of either of these growth factors. As shown by McAuslan, Reilly, Hannan and Gole (522) intense corneal neovascularization is reproducibly induced in the corneal pocket assay with Elvax polymers impregnated with 2×10^{-11} M formyl methionyl leucyl phenylalanine, but not when this compound was administered at concentrations of 10^{-10} M or greater or at 10^{-12} M or lower. BenEzra (55) obtained no vascularization with this synthetic peptide in 23 assays with 20 µg per Elvax implant. With some substances, such as heparin and adenosine diphosphate (ADP) (522), the corneal angiogenic response is variable suggesting a dose dependency. The absence of an induced neovascular response may reflect the lack of a direct or indirect angiogenic factor; alternatively too low a concentration of the stimulus may fail to elicit a response. When high concentrations of compounds, such as adenosine diphosphate, fail to evoke the angiogenesis that lower concentrations elicit suggests an inhibitory effect at higher concentrations (522). Thus far most reports of putative substances that do not elicit corneal neovascularization have not taken the aforementioned into account.

Chapter 6

SUPPRESSION AND ENHANCEMENT OF CORNEAL ANGIOGENESIS

Valuable insights into the pathogenesis of neovascularization in the cornea are gained by an understanding of circumstances that suppress and enhance angiogenesis in this tissue. Studies in a variety of experimental models have demonstrated a suppression of corneal neovascularization in leukopenic animals. Corneal angiogenesis induced by chemical cautery is severely suppressed under situations that inhibit the inflammatory response and most notably the leukocytic infiltration into the cornea. As described below, this occurs following treatment with corticosteroids, non-steroidal anti-inflammatory drugs and irradiation (head irradiation, total body irradiation, total lymphoid irradiation, and total body irradiation followed by bone marrow transplantation) Together these findings add further support to the hypothesis that corneal vascularization is a component of inflammation and that leukocytes play a cardinal role.

SUPPRESSION OF ANGIOGENESIS

1. Steroids

Since Ashton, Cook and Langham drew attention to the anti-angiogenic effect of topical corticosteroids in 1951 (20), numerous investigators have established beyond all doubt that topically administered corticosteroids suppress, but do not completely inhibit, corneal vascularization in a variety of clinical and experimental circumstances (128,155,264,370,416,515,533,535,572,586,677,806). This suppression of corneal neovascularization has been noted with different corticosteroids including cortisone (20,469,572,816), triamcinolone (533), methylprednisolone (264,515), dexamethasone (128,333,337,392,494,533,593,776), prednisolone (128,155,337,354, 494,533,597,636), and ticabesone propionate (494). Indeed in some studies the timing of the drug administration seems to effect the degree of angiogenesis with the most suppression of new vessel growth occurring with treatment prior to, or immediately after, corneal injury rather than after the onset of the lesion (128,264,515,636). For example, pretreatment of rats with corticosteroid markedly depresses the corneal angiogenic response to intracorneal copper, Dispirin® and Walker carcinoma extracts (515). Subconjunctival methylprednisolone acetate administered to rats immediately after or 24 hours after corneal injury with silver nitrate inhibited the infiltration of leukocytes and the subsequent vascular invasion of the corneal stroma if administered immediately after cauterization. On the other hand, it did not prevent the invasion of the cornea by blood vessels if it was instilled 1 day after cauterization, at which time leukocytes had already infiltrated the cornea. However, under such circumstances the corneal leukocytic and vascular ingrowth was less severe than in the nonglucocorticoid-treated corneas (264). Not all investigators have reported success in the suppression of corneal neovascularization with corticosteroids (159,451,459) and it is noteworthy that these studies have all used the alkali-burned cornea as a model of corneal angiogenesis.

The suppressive effect of corticosteroids on angiogenesis is not specific to the cornea and has been observed, for example, in the endothelial proliferation around a freeze injury to the cerebral cortex (51), and in the granulation tissue of healing gastric ulcers (334).

How corticosteroids suppress the growth of blood vessels into the cornea remains incompletely understood, but this suppression presumably relates at least in part to the local anti-inflammatory action of the steroids (832). Part of the effect could be due to the inhibition that corticosteroids have on the migration of leukocytes towards sites of injury in the cornea, as pointed out more than a decade ago by Fromer and Klintworth (264) and confirmed by others (128,597,636). However, the situation is clearly more complex as glucocorticosteroids also inhibit the dilatation and increased permeability of blood vessels to polymorphonuclear leukocytes (84), cause a marked reduction in the macrophage population of some lesions (51), destroy lymphocytes (137) and diminish other participants of the inflammatory response. Corticosteroids stabilize lysosomal membranes (797) and hence inhibit the release of hydrolytic enzymes from polymorphonuclear leukocytes, diminish the increased vascular permeability that accompanies inflammation (732), inhibit the production of several cytokines including interleukin 1 (IL-1) (64,710) and tumor necrosis factor (TNF) (67). Moreover, systemic corticosteroids induce monocytopenia (457). Corticosteroids also effect the vascular endothelium directly. For example they can inhibit prostacyclin synthesis by endothelial cells (387). Hydrocortisone also has a slight inhibitory effect on DNA synthesis and cell growth of certain vascular endothelial cells (mouse cerebral microvasculature) (51).

The anti-inflammatory effect of corticosteroids has been shown to result from the binding of the steroid to a cytoplasmic receptor, followed by the translocation of this complex to the nucleus where the synthesis of lipocortin, a "second messenger" is induced. This protein prevents the activation of phospholipase A_2 (E.C.3.1.1.4) (352) the enzyme that hydrolyzes phospholipids to nonesterified fatty acids. Blockage of phospholipase-A_2 activity interfers with the liberation of arachidonic acid from cell membranes and the synthesis of the entire cascade of eicosanoids (prostaglandins, leukotrienes, HETEs and other metabolites of arachidonic acid) by the cyclooxygenase and lipooxygenase pathways (228).

Heparin combined with cortisone enhances the anti-angiogenic effect of cortisone on post-traumatic (572), alkali-induced (721) and prostaglandin E_1 (PGE$_1$)-induced (128) corneal vascularization. The reason for the synergism between heparin and corticosteroids remains unclear, but the combination of these two compounds also inhibits the growth of new capillary blood vessels in the chick embryo, in the rabbit cornea following tumor implantation as well as in some mouse tumors (244). Cycloamyloses, which act as carriers for hydrophobic molecules, such as certain steroids, and can adsorb to endothelial cells have an even stronger synergism with steroids (cortexolone) against endotoxin induced corneal neovascularization and angiogenesis induced on the chick chorioallantoic membrane (247). Folkman et al. (247) suggest that the steroid/heparin synergism may also be due to the "carrier molecule" property of heparin. Steroids that lack glucocorticoid or mineralocorticoid activity also inhibit angiogenesis in the chick chorioallantoic membrane in the presence of heparin or fragments of it (161). Whereas corticosteroids clearly effect corneal neovascularization, cortisone alone is stated to not inhibit tumor angiogenesis except in the presence of heparin (244).

Medroxyprogesterone decreases substantially the polymorphonuclear leukocytic and vascular infiltration into sensitized rabbit corneas with experimental herpes simplex keratitis (450). It also decreases moderately corneal neovascularization following thermal burns (597) or PGE$_1$ induction (128). This angiostatic steroid has

17

been shown to inhibit plasminogen activator, an enzyme implicated in angiogenesis, in bovine endothelial cells (14).

2. Non-steroidal Anti-Inflammatory Drugs

Non-steroidal anti-inflammatory agents reported to retard corneal neovascularization include inhibitors of fatty acid cyclooxygenase (also known as prostaglandin synthetase) (E.C.1.14.99.1): flurbiprofen (155,183,337,401,539,593), indomethacin (174,266,333,337,636,700,829) and Ketorolac (337,494). While indomethacin does not abolish the ability of prostaglandin to trigger blood vessel proliferation, it appears to limit the perpetuation of vascularization as well as its extent (55). Indomethacin is reported to suppress corneal vascularization induced by alkali burns by some investigators (266), but not by others (159,776). Indomethacin is also reported to not affect corneal angiogenesis induced by the experimental denudation of the corneal epithelium with a motor-driven rotating stone (716). Corneas treated with indomethacin drops contain less PGE_2 following alkali burns than corneas with similar injuries, but without indomethacin treatment. Why different investigators should obtain a variable effect of indomethacin on corneal neovascularization remains unclear, but it may reflect at least in part the experimental model, the method used for quantitating angiogenesis, and/or the animal species used in the study. In addition, topically administered sodium salicylate, an inhibitor of prostaglandin synthesis, is stated to not effect angiogenesis induced by a thermal burn of the cornea (354).

Inhibitors of lipooxygenase (REV 5901, esculetin and quercetin) have not been effective in suppressing experimentally, induced corneal vascularization (275). Phenidone, a dual inhibitor of lipooxygenase and cyclooxygenase, suppresses corneal neovascularization following silver/potassium nitrate cauterization, but since this compound is a photographic developer the suppression of angiogenesis may reflect the reduction of silver rather than a direct effect on the metabolic pathway leading to new vessel formation (337,494). For reasons that remain to be determined other dual cyclooxygenase/lipooxygenase inhibitors (BW 755C, BW A540C) do not reduce corneal neovascularization induced by silver/potassium nitrate cautery in the rat (337).

3. Irradiation

Corneal angiogenesis is severely suppressed by total body irradiation (194,264,290,654,694) and total lymphoid irradiation (731). It is less severely suppressed by head irradiation (290) and by total body irradiation followed by bone marrow transplantation (290).

Fromer and Klintworth (264) observed that neither leukocytes nor blood vessels invaded the cauterized corneas of weanling rats whose circulating leukocytes were totally eliminated by massive doses of total body irradiation; whereas, both a leukocytic and vascular invasion occurred at lower doses of irradiation that did not totally eliminate circulating leukocytes. The degree of resultant corneal vascularization was dependent upon the timing of the irradiation in respect to the corneal cauterization. The corneas vascularized if they were cauterized immediately after total body x-irradiation; that is, prior to the onset of leukopenia. When the cornea was cauterized four days after x-irradiation limited to the head, corneal angiogenesis still ensued. Corneas of rats cauterized after only their heads received a similar dosage of x-irradiation ruled out the possibility of irradiation-induced limbal vascular endothelial damage as the explanation for the vascular suppression observed by x-ray treatment. Subsequent studies under different experimental conditions have shown that although corneal vascularization follows irradiation of the head the amount is less than in non-irradiated animals (290,695), presumably because the blood vessels that invade the cornea in those cases are formed only from endothelial cell migration rather than in the usual manner by migration and mitosis. In the studies by Sholley, Gimbrone and Cotran (694) the leukopenic effect of total body irradiation was enhanced by antineutrophil serum. Current evidence suggests that the suppression of corneal vascularization by total body irradiation is due to the effects of irradiation on the peripheral leukocyte count and on its inhibition of the pericorneal microvasculature's ability to undergo cell division, as well as by a yet to be defined nonspecific effect of irradiation on other tissues. In quantitative studies of corneal vascularization in inbred mice following chemical cautery (silver/potassium nitrate), the degree of angiogenic suppression produced by total body irradiation is significantly less if a total marrow transplant is performed immediately after the irradiation supporting the belief that bone marrow derived elements are involved in corneal angiogenesis in that model (290). However, since the degree of corneal neovascularization in these reconstituted mice was not equivalent to those that only received head irradiation suggested that total body irradiation may inhibit corneal neovascularization by not only the radiation effect on the pericorneal vasculature and the bone marrow, but also by some yet to be defined additional effect. Whereas beta irradiation has not been shown to prevent the onset of experimentally induced corneal vascularization it is reported to be effective in enhancing the regression of experimentally induced corneal neovascularization (370,698). Corneal angiogenesis has also been found to be diminished in leukopenic rabbits following total body irradiation with eyes shielded (195).

In rats, a single dose of 30 Gy severely inhibits angiogenesis if delivered prior to wounding in the Selye pouch (775). Sholley et al. (692) have shown that vascular sprouting is not dependent upon endothelial cell division in the silver nitrate cauterization model of corneal vascularization. X-irradiation (2000 or 8000 rads) to the eye prior to cauterization impairs endothelial cell proliferation, but vascular sprouting is similar to non-irradiated corneas. However, by 4 days the vascular ingrowth is reduced 66.7 and 53.4% of control values after 2000 and 8000 rads, respectively (692).

4.   Protamine

Protamine, an arginine-rich basic protein that binds heparin, blocks the ability of mast cells and heparin to stimulate the migration of capillary endothelial cells (34). It also suppresses the rate of corneal angiogenesis in rabbits after the intracorneal implantation of silica particles or allogeneic lymph nodes (748). Taylor and Folkman (748) found that polymer pellets of ethylene vinyl acetate (EVA) containing protamine positioned between a tumor (V2 carcinoma) implanted in the rabbit cornea and the corneoscleral limbus suppressed corneal angiogenesis, and that the corneal capillary growth resumed before tumor growth after the removal of the protamine.

5.   Antimitotics

Triethylene thiophosphoramide (Thiotepa) blocks the multiplication of capillary endothelial and other cells and is effective in suppressing corneal vascularization (370,447,451), but it has toxic side effects (370).

6.   Tissue Extracts

(i)   Cartilage Extracts: Inhibitors of angiogenesis have been isolated from cartilage (93,193,432,442,444,454,554,712) and cells grown from mouse, rat and rabbit costal cartilage produce factors that suppress vascular endothelial cell growth (740,741). Intracorneal implants of hyaline cartilage, which contains anti-angiogenic material (193), as well as an extract of cartilage containing antiprotease activity inhibit tumor induced capillary proliferation in the rabbit cornea (93,442). Brem and Folkman (93) observed that a tiny piece of neonatal cartilage (but not boiled cartilage) placed in the corneal stroma between the limbus and the tumor explant prevented some tumors (28%) from vascularizing. A protein, which inhibits mammalian collagenase, has been isolated from cartilage and found to inhibit capillary endothelial cell proliferation in culture and angiogenesis in the CAM assay (554).

(ii)   Vitreous: Vitreous and extracts of it from several species (rabbit, human, bovine) inhibit angiogenesis in a variety of bioassays (94,95,192,218,221,482,527,528), including neovascularization induced by tumor implants in the rabbit cornea (611). Vitreous from bovine, human and chick embryo has been found to contain fractions that inhibit aortic endothelial cell growth activity (379). Hyaluronate, a major constituent of the vitreous, may contribute to this anti-angiogenic effect, since hyaluronate inhibits or reduces the growth of blood vessels in several systems including chick embryo limb buds (217) and in granulation tissue around subcutaneous foreign body implants in rats and guinea pigs (41,653). Some investigators have not detected a significant anti-angiogenic effect of hyaluronic acid in some assay systems (482) and new vessel growth has been noted close to sites that are normally rich in hyaluronate, as in rheumatoid arthritis, osteoarthritis and diabetic retinopathy. If hyaluronate inhibits angiogenesis the mechanism remains incompletely understood. This polyanionic macromolecule inhibits the movement of fibroblasts and several other cell types (including mononuclear phagocytes, granulocytes, and lymphocytes) (41), which have been implicated in corneal neovascularization, as discussed elsewhere in this review. This motility inhibition is proportional to the concentration of the hyaluronate, and it is abolished when this glycosaminoglycan becomes degraded by hyaluronidase or by oxidation-reduction systems (41). Depending upon the concentration and size of the hyaluronic acid molecule, hyaluronate also inhibits the ability of lectins and other mitogens to induce lymphoblast formation in cultured lymphocytes (41), and such involvement may inhibit lymphocyte induced angiogenesis. Aside from evidence implicating hyaluronan in the inhibition of angiogenesis, degradation products of this glycosaminoglycan have been implicated in angiogenesis. Angiogenic activity has been noted in the chick chorioallantoic membrane in response to a bovine testicular hyaluronidase digest of human umbilical vein hyaluronan (12). Angiogenic activity has been attributed to hyaluronate fragments (4-25 disaccharides in length) using the chick chorioallantoic membrane assay (801), and an increased number of small vessels has been noted in ischemic myocardium after treatment with high doses of hyaluronidase (421). Also, the vascular ingrowth into some tissues is associated with tissue hyaluronidase activity and a decrease in the hyaluronic acid concentration (53,764).

Constituents of the vitreous other than hyaluronic acid may also contribute to its anti-angiogenic as some vitreal extracts with putative antiangiogenic activity have been insensitive to enzymes like hyaluronidase and chondroitin sulfate lyases that degrade this glycosaminoglycan (221,482).

(iii)   Aortic Extracts:

At least four substances extracted from intact aorta inhibit endothelial cell growth in culture and smooth muscle cells derived from the aorta produce such a factor (676). When administered either subconjunctively or topically a low molecular weight extract of bovine aorta inhibits corneal angiogenesis and enhances the regression of newly formed corneal blood vessels in rabbits (192).

(iv)   Cornea: Corneal neovascularization has been inhibited by a proteinase inhibitor extracted from cornea (431).

7.   Other

In studies involving a variety of corneal disorders and the lack of appropriate controls and objectivity the subconjunctival injection of 100% oxygen is reported to cause the regression of corneal vascularization in human subjects (583,584).   The instillation into the conjunctival sac of preparations of certain unsaturated fatty acids (columbinic acid, eicosapentaenoic acid, γ-linolenic acid dihomo-γ-linolenic acid) capable of interacting with cyclooxygenase is reported to suppress corneal angiogenesis induced in the rabbit by the intracorneal injection of human serum albumin (777). The vascularity of corneal grafts is reduced by cyclosporine A in the rat after systemic administration (345), but not in rabbit following the topical delivery of this drug (806). Other substances with a putative anti-angiogenic effect, but which have not been tested on corneal angiogenesis, include gold compounds (503).

ENHANCEMENT OF CORNEAL VASCULARIZATION

The amount of corneal angiogenesis induced by a focal corneal lesion is related to the distance of the injury from the corneoscleral limbus (116), as well as by the size of the lesion. In several quantitative studies of corneal angiogenesis in the rat following silver/potassium nitrate cauterization using computerized image analysis the degree of corneal vascularization has been directly proportional to the area of the cautery site (647,726,731).

Studies using computerized image analysis of carbon filled corneal blood vessels have shown that the corneal angiogenic response to chemical cautery with silver nitrate/potassium nitrate is enhanced in nude mice (703), mice immunized with rabbit anti-mouse-platelet serum (302), and in rats whose eyelids have been sutured or patched closed after corneal cautery (especially if preceded by a retrobulbar injection of saline, epinephrine or epinephrine plus local anesthetic) (647). The reason for the angiogenic enhancement in these situations remains unclear, but future studies in these animal models may shed light on the phenomenon.

Campbell and Ferguson (115) found that corneal vascularization could be more frequently elicited in scorbutic guinea pigs than normal guinea pigs after injury to cornea.

The addition of gangliosides (mono- and trisialotetraesosyl ganglioside sodium salts) to prostaglandin E_1 or basic fibroblast growth factor has been reported to increase the frequency of corneal neovascularization following its introduction into the cornea (827), but the induced angiogenesis under these conditions has not been quantitated satisfactorily for adequate statistical analysis.

Chapter 7

HYPOTHESES ABOUT THE PATHOGENESIS OF CORNEAL VASCULARIZATION

Many factors influence capillary growth and numerous theories have been proposed to account for the formation of new capillaries in different situations (114,649).

Destruction of Anti-Angiogenic Substance

For theoretical reasons the normal cornea may contain at least one inhibitor of angiogenesis, but the question of whether this is true still awaits a definitive answer. Anti-angiogenic factors could be responsible for the cornea's avascularity and if present normally their loss would enhance the growth of blood vessels into the cornea. Chemical similarities between the cornea and other avascular tissues (35) originally prompted consideration of this possibility. Meyer and Chaffee (532), impressed by the abundance of mucopolysaccharides (currently termed glycosaminoglycans) in the cornea and other avascular tissues, such as cartilage and Wharton's jelly, proposed that these compounds might inhibit the growth of blood vessels into the cornea. They also found that the intracorneal instillation of hyaluronidase induced corneal vascularization, whereas the injection of inactivated enzyme usually provoked only a transient reaction (532). Michaelson, Herz and Rapoport (537), however, did not find daily subconjunctival injections of hyaluronidase to influence corneal vascularization induced by electrocautery. If glycosaminoglycans prevent tissue vascularity, one would expect differences between the glycosaminoglycan content of normal and vascularized corneas. In an early study non-vascularized and vascularized corneas (induced by riboflavin deficiency) were found to contain equivalent amounts of glucosamine, a constituent of hyaluronic acid (810). By histochemical methods an apparent diminution in corneal glycosaminoglycan content has not been observed prior to vascularization, or in the region in front of the invading blood vessels (16) and no conspicuous abnormality has been noted in the existing ^{35}S-sulfate labeled corneal glycosaminoglycans during neovascularization (705,706). For these reasons the theory of a natural inhibitor of corneal vascularization fell into disrepute. A critical review of the data, which led to the downfall of the theory, reveals many loopholes. Normally glycosaminoglycans are covalently bound to proteins as proteoglycans and the main glycosaminoglycan component of the cornea, keratan sulfate, is not degraded by hyaluronidase. One yet to be confirmed experimental study, however, suggests that corneal fibroblasts may inhibit lymphocyte induced angiogenesis by an interaction with these leukocytes. In this investigation isolated corneal cells decreased the angiogenic response evoked by the intradermal injection of semi-allogeneic lymphocytes (397). The claim that glycosaminoglycans do not inhibit corneal angiogenesis has also not been adequately tested experimentally. The techniques used by Ashton (16) as well as Smelser and Ozanics (705,706) are not as sophisticated or as precise as those that can now be applied to the problem.

Although yet to be demonstrated with assurance, the possibility of the normal cornea containing vasoinhibitory substances still warrants serious consideration. Certain avascular tissues have long been suspected of harboring anti-angiogenic substances and there is supporting experimental evidence for angiogenic inhibitors in cartilage (93,193,432,442,444,454,712), the aorta (192,676), and vitreous (93-95,192,218,221,482,527,528,611). Moreover, in a recent report a proteinase inhibitor extracted from the cornea has been found to suppress corneal angiogenesis (431).

Also, noteworthy is the observation that corneal tissue with macular dystrophy and the systemic mucopolysaccharidoses (characterized by an excessive corneal accumulation of material with the cytochemical attributes of glycosaminoglycans) is usually avascular.

In 1947 Bachsich and Wyburn (36) proposed that corneal neovascularization was a sequel to the destruction of an anti-angiogenic factor which normally restrained neighboring vascular cells from invading the cornea.

Corneal Edema Eliminating Mechanical Barrier

In 1948 Mann, Pirie and Pullinger (496) made the important observation that blood vessels only invade the corneal stroma when the tissue is edematous. A year later Cogan (143) hypothesized that the edema was cardinal in the induction of the angiogenesis as it loosened the corneal stroma thereby permitting the ingrowth of vessels that had been restrained by the compact collagen bundles in the normal cornea. This theory was partly based on the observation that some avascular tissues, such as fingernails and cartilage, offer a mechanical barrier to blood vessels and the clinical finding that corneal edema consistently accompanies and frequently precedes corneal vascularization.

Corneal edema seems to precede vascularization in most, if not all, situations that lead to new vessel growth in the cornea and a loosening of the corneal stroma undoubtedly facilitates the invasion of this tissue by blood vessels. However, several observations and theoretical considerations argue against the view that corneal edema elicits the angiogenesis. Ashton and Cook (18) stressed that vascular endothelial cells need a stimulus to proliferate and that this is unlikely to be in response to a reduction in tissue compactness. Also, because corneal vascularization is not an inevitable sequel to corneal edema, it is not sufficient to provoke angiogenesis (229,359,416,446,463,506,512). Additional evidence against edema being a stimulus for corneal vascularization includes: (i) Baum and Martola's (48) finding by pachometry that corneas with vascularization were thinner than nonvascularized corneas with bullous keratopathy, (ii) the degree of corneal vascularization is not directly related to the amount of stromal edema, (iii) corneal tissue does not necessarily vascularize when transplanted into hamster cheek pouch chambers, even when corneal lamellae are separated in a swollen stroma (416), (iv) during development in the chick embryo the corneal stroma becomes edematous (157), but this is not followed by vascularization of the swollen tissue, (v) corneal neovascularization is rare in some settings in which corneal edema is prominent, such as Fuchs' dystrophy (182,419,506), congenital endothelial corneal dystrophy (506), bullous keratopathy (48) and other disorders of the corneal endothelium, such as after filling the anterior chamber with silicone (512) or after the experimental destruction of corneal endothelial cells with intracameral benzalkonium chloride in the rabbit (511).

The loosening of the stromal framework that corneal edema causes undoubtedly provides capillaries with space into which they may extend and for ample reasons one would suspect that vessels will grow preferentially in planes of diminished resistance. Evidence that swelling of the cornea facilitates the ingrowth of blood vessels is found in experiments in which corneal buttons freed of the epithelium and endothelium were fixed between glass filter discs and placed subcutaneously into the rabbit from whom they were excised allowing each corneal stroma to swell according to the space between the apposing discs. The corneal stroma was richly vascularized in corneas with swollen tissue (1.0 mm) but not in those kept at normal thickness (0.35 mm) (182).

Production of Angiogenic Factor in Cornea

The possibility of a locally generated factor(s) triggering corneal angiogenesis has long been suspected and an increasing body of evidence supports this view. When a localized corneal abnormality provokes angiogenesis, in a clinical or experimental setting, the new blood vessel formation usually begins at the corneoscleral limbus nearest to the lesion, and the newly formed capillaries often extend into the peripheral cornea within an isosceles triangle that has its base at the corneal periphery (116). Moreover, as pointed out by Campbell and Michaelson more than four decades ago (116), a localized injury typically elicits an angiogenic response only if it is situated within a critical distance of the corneoscleral limbus. Such observations can be explained by a factor produced within the corneal lesion and which diffuses centripetally as its concentration diminishes until it becomes ineffective beyond a critical distance from the point of injury. In 1966 Maurice, Zauberman and Michaelson (512) provided additional evidence for the existence of an angiogenic factor in corneal neovascularization. By repeatedly wounding the tissue close to the central end they were able to cause vessels to grow into open plastic tubes that had been implanted into corneas suggesting that an angiogenic factor liberated at the wound passed down the lumen of the tube to the vessel walls. By assuming values for the permeability and diffusion constants of the angiogenic factor based on the diffusion rate of fluorescein within the rabbit cornea and the postulate that the angiogenic factor is an oligopeptide with a molecular diameter perhaps half again as great as that of fluorescein, Maurice, Zauberman and Michaelson estimated its distribution in the tissue and its range of action (512).

Several investigators have alluded to the presence of angiogenic factors in the potentially vascularized (but not normal) cornea (18,116,229,263-265,283,374,416, 417,512,822). In hamster cheek pouches, newly formed capillaries commonly extend toward and into corneal explants, consistent with the hypothesis that specific substances possess the ability to stimulate vascular growth (416). Gimbrone, Cotran, Leapman and Folkman (283) also stressed that when tumors are implanted in the corneal stroma, the time taken to vascularize is directly proportional to the distance between the tumor and the limbus. Implants in the central cornea (4 to 5 mm from the limbus) sometimes remain unvascularized for up to 2 months. Like other theories, this one has also had critics. For example, Cogan (143) observed that a small injection of sodium hydroxide produced a swollen zone around the site of injection and that the blood vessels grew into the edematous region rather than into the injured area. However, this inconsistent finding (512) does not carry much weight against the belief in an angiogenic factor. As pointed out by Ashton (16), vessels do not grow into necrotic tissue. Also, the source of the angiogenic factor may be distant from injury and not a specific product of the injured tissue. Some investigators have questioned the angiogenic factor theory of corneal vascularization because the newly formed vessels cease migrating beyond a certain point and do not grow beyond the corneoscleral limbus (18). Objections on this ground are clearly flawed as an angiogenic factor diffusing away from the cornea could drain away from the eye without reaching a sufficient concentration to provoke an effect on neighboring conjunctival and scleral blood vessels.

Hypoxia

New blood vessels form from the microvasculature in several apparently different normal and pathologic situations. Angiogenesis is a feature of cell replication (as in metazoa in which a circulatory system evolved of necessity as a functional adaptation, embryonic development, and solid neoplasms), inflammation and tissue repair (as in wound healing by granulation tissue), and tissue hypoxia (as in retinal neovascularization). All the aforementioned have in common an increased cellularity or tissue hypoxia followed by vascularization. One would expect vast quantities of cells in close proximity to each other to compete for oxygen and existing nutrients and for certain metabolites to accumulate and alter the cellular milieu and possibly cause the cells to liberate some factor capable of inducing directional vascular growth. Since the normal corneal stroma is relatively acellular, could it be that an increased cellularity contributes to the invasion of the cornea by blood vessels?

In certain situations angiogenesis follows tissue hypoxia and a low oxygen tension is suspected of being important in the neovascularization. These settings include vascularizing wounds where the oxygen tension is invariably low in the center of the lesion (425,627,699), as well as proliferative diabetic retinopathy and the retinopathy of prematurity (15,17).

Several observations have led investigators to suspect an angiogenic product of anaerobic metabolism in the pathogenesis of corneal neovascularization: corneas of rats deprived of riboflavin, a coenzyme of several respiratory enzymes, become vascularized (65,90,189,263,810); the corneal vascularization in some humans with rosacea keratitis responds to riboflavin treatment (386). Because vascularizing corneal tissue is analogous to the granulation tissue of wound healing one would suspect a similar situation in cornea. Moreover, the cornea is dependent upon the aerobic metabolism of carbohydrate for its energy needs and this requires atmospheric oxygen and takes place mainly in the epithelium (445). Some corneal contact lenses form a barrier to this source of oxygen and a complication of their use is corneal neovascularization (80,183,470,568,720). The possibility of lactate, a product of anerobic metabolism, playing a role in corneal angiogenesis has been raised and the intracorneal injection of lactic acid elicits a vascular invasion of this tissue, albeit inconsistently (374). In normal rabbits the lactic acid concentration of the central cornea is greater than that of the peripheral cornea and the lactate levels in the peripheral cornea have been found to increase to the level found in the central cornea (evidence of anaerobic glycolysis) prior to corneal neovascularization induced by an ocular encircling rubber band (463). Lactate does not cause an angiogenic response on the chick chorioallantoic membrane (255), but Jensen, Hunt, Scheuenstahl and Banda (385) have observed that lactate stimulates macrophages to secrete factors with angiogenic activity using rabbit corneal implants as the assay system.

Based on observations in a study of rabbits with silicone-filled anterior chambers, Maurice, Zauberman and Michaelson (512) maintained that hypoxia (or suboxidation) could not constitute a stimulus to corneal vascularization as "the state of oxidation would have to be assumed to increase when vascularization ceased - 2 weeks after the experiment started, and to then drop again when the cornea was wounded, but only in the segment in which the wound was sited". This argument, however, is negated if the oxygen drop is caused by oxygen utilization by leukocytes which infiltrate the injured region.

The influence of oxygen tension in corneal angiogenesis has been evaluated experimentally in several models. Lazar, Lieberman and Leopold (452) observed that corneal vascularization induced in the rabbit by alkali burns was not inhibited by 100% oxygen (8 hours/day x 4 days; 2 atmospheres) (452). Moreover, continuous exposure of rabbits to elevated oxygen levels for 6-7 days following intracameral alloxan or thermal burns do not prevent corneal vascularization (19,536) and nor has experimentally produced corneal vascularization been prevented in the rabbit (393) and guinea pig by exposure to hyperbaric oxygen (342). However, these evaluations of the potential influence of hypoxia in the pathogenesis of corneal neovascularization were not performed under ideal conditions in an objective non-biased manner on an adequate number of animals in different experimental models. In a study in which known concentrations of oxygen (0%, 10%, 21%, 50%, 75% or 100%) were perfused through goggles fitted over both eyes of the rat following chemical cauterization of the cornea, we recently measured the resultant neovascularization of India ink filled vessels in flat preparations of corneas quantitatively using computerized image analysis (162). The angiogenic response of rats whose eyes were continuously exposed to 0-75% oxygen were not significantly different from each other. The mean response to 100% oxygen was, however, 10-21% lower than all of the other groups. The reason for the inhibitory effect of 100% oxygen remains to be determined but it may represent a toxic effect of oxygen free radicals on the vascular endothelium.

Despite the likelihood that tissue hypoxia contributes to contact lens induced neovascularization, it may, like other types of corneal vascularization, also be secondary to the inflammatory response, because corneal angiogenesis induced in rabbits with extended wear contact lenses is significantly suppressed with the anti-inflammatory agent flurbiprofen (a cyclooxygenase inhibitor) (183).

Manifestation of Inflammation

Sequential morphological observations on heterotopic corneal grafts (from various sources and after various forms of pretreatment) in the hamster cheek pouch first led me to suspect that corneal neovascularization was a manifestation of the inflammatory response and that leukocytes might play an important pathogenetic role. Several years ago I inserted a variety of normal, injured, and nonviable corneas into hamster cheek pouch chambers and unexpectedly found that the vascularization of the transplanted tissues was virtually independent of the nature of the graft (416). The crucial factor appeared to be the host's reaction to the transplanted corneal tissue. Unless it provoked an inflammatory cell infiltrate, corneal vascularization did not occur. Since then a review of experimental models of corneal vascularization has revealed that the inflammatory reaction occurs in most of them prior to neovascularization (263,417). In all situations in which corneal vascularization has been thoroughly studied it has occurred in association with the inflammatory response and the degree of angiogenesis has often been proportional to the vigor of the inflammatory reaction. Numerous studies emphasize the association of corneal vascularization with the inflammatory response (416,822) and most facets of corneal neovascularization are identical to the angiogenesis of inflammation as occurs in other tissues.

The hypothesis that corneal angiogenesis is a manifestation of inflammation implies that the extent and nature of the inciting agent as well as the degree of inflammation may influence the degree and duration of the neovascularization. That this is true is supported by several observations: (i) capillaries invade the cornea after leukocytes in a wide variety of situations, including corneal injuries produced by chemicals (silver nitrate, sodium hydroxide, colchicine, or alloxan), antigens

27

(intracorneal antigens in sensitized animals), and metabolic disorders (hypertyrosinemia, as well as riboflavin and vitamin A deficiency) (263-265,419), (ii) although the lesions in different experimental models of corneal neovascularization vary all models apparently display three phases: an early prevascular phase of leukocytic infiltration, a second phase where leukocytes and blood vessels occur together and a later phase where a corneal vasculature persists in the absence of leukocytes (263), (iii) the prevascular latent period is directly related to the time of the initial leukocytic infiltration (263), (iv) in different experimental models new vessels grow into the cornea only in areas in which leukocytes are present (263).

Although angiogenesis clearly can occur in the absence of inflammation in some biological settings, such as during development, an exceptable model of non-inflammatory corneal neovascularization has still to be found despite a long time search for one. Alleged non-inflammatory models have turned out not to be valid on closer scrutiny. Blood vessels are claimed to invade the cornea in the absence of leukocytes after neoplastic cells are instilled into the cornea (283). Others have also claimed to induce corneal angiogenesis in the absence of an inflammatory reaction with PGE_1 (829), and extracts of tumor (7). Also, a problem in interpreting the early events in corneal angiogenesis stems from the observation that once the cornea becomes vascularized, the blood vessels may remain throughout life. They are frequently devoid of blood, but their presence can be detected by slit-lamp examination and is a testimony to the previous seat of corneal inflammation. Such vessels may loose their circulation but rapidly regain it as part of the hyperemic response to a subsequent corneal injury. Because of this, studies concerned with the pathogenesis in corneal vascularization need to exclude the presence of existing corneal capillaries in experimental animals. To date authors of putative non-inflammatory corneal neovascularization have not provided convincing evidence that the inflammatory response was not present. Claims of corneal angiogenesis in the absence of inflammation have been made (7,154,215,306 and others). In some of these instances, I have personally studied the material and invariably found a prominent accompanying leukocytic infiltrate which the authors have failed to observe. When angiogenesis is believed to be specifically induced it is essential that an associated inflammatory response be excluded. This requires that microscopic evaluations of the cornea be performed after the instillation of the putative angiogenic factor at appropriate times prior to and during the early neovascularization. The mere documentation of the absence of an associated leukocytic infiltrate in corneas does exclude inflammation induced angiogenesis since blood vessels persist within the cornea long after this hallmark of inflammation has subsided (263). Moreover, even if a leukocytic infiltrate is excluded as in irradiated leukopenic animals, the presence of non-cellular constituents of the inflammatory response need to be ruled out as well since they also apparently contribute to the angiogenesis of inflammation. Some claims (7), have been based on clinical observations, which do not provide adequate magnification, and without sufficient histopathologic documentation at sufficient time periods between the initiation of the corneal injury and the onset of the angiogenic response. Many investigators have only performed histopathologic evaluations when the vascularization was flagrant at which time the antecedent inflammatory cell infiltrate may have subsided; others have suppressed the leukocytic infiltrate and, while not obliterating the humoral aspects of inflammation, have assumed incorrectly that the inflammatory response was not present because of the absence of a leukocytic infiltrate. Because he was able to produce corneal angiogenesis in irradiated rabbits (700 rads whole-body irradiation) by thermal cauterization in the apparent absence of invading leukocytes Eliason (194) stated "it can be concluded only that, in the absence of invading leukocytes, an injured cornea is capable of producing an angiogenic factor from its native elements". This conclusion, however, totally ignores the possible role of the non-cellular components of inflammation in the angiogenic response. Surprisingly, some investigators have claimed that inflammation was not present in scientific presentations or publications, but have illustrated intracorneal leukocytes in their figures (306)! As in the past I throw down the gauntlet to any investigator who can find a bona fide experimental model of non-inflammatory corneal angiogenesis.

Release of Sequestered Heparin-Binding Growth Factors

Recently Folkman's group of investigators have suggested that Descemet's membrane, and to a lesser extent other parts of the cornea, contain sequestered angiogenic factors, such as heparin-binding growth factors, for capillary endothelial cells and that the abnormal release of these factors could be responsible for a variety of different types of corneal neovascularization (242,781).

Chapter 8

POTENTIAL SOURCES OF ANGIOGENIC FACTORS

Several investigators have focused on the source of the angiogenic activity and progress is only now beginning in the assessment of the relative roles of the many potential participants in corneal angiogenesis. In theory the angiogenic factor(s) responsible for corneal vascularization could arise from one or more of the following sources: the denatured tissue, the injurious agent, corneal cells, cells of noncorneal origin, pericorneal vascular cells, serum, tears, or the aqueous humor.

A.   Components of the Inflammatory Response

Within the complex cellular and humoral events of the inflammatory response many potential sources exist for the initiation and potentiation of corneal angiogenesis.

1.   Cellular Components

Seventeen years ago in studies on corneas implanted into hamster cheek pouch chambers, I was especially inspired by the observation that the invasion of the cornea by blood vessels was invariably accompanied by a cellular infiltration of the corneal stroma. When the vascular invasion involved only part of the cornea, the blood vessels were at the site of the cellular infiltrate. If cells did not penetrate the corneal tissue under investigation, it consistently remained avascular. As a result of these observations I hypothesized that leukocytes may be a prerequisite to corneal vascularization and that under certain circumstances they may produce one or more factors capable of stimulating directional vascular growth (416). This view was reinforced in subsequent experiments performed with Fromer (263-265) and much data supports the notion that leukocytes produce one or more factors which stimulate directional vascular growth (263-265,416-418,515,636,637,664). That leukocytes play a role in angiogenesis is supported by the following observations: (i) the events that precede and accompany corneal vascularization in several apparently diverse established experimental models of corneal vascularization in rats and rabbits are essentially similar and in almost all models of corneal vascularization that have been thoroughly studied, leukocytes have invaded the cornea before blood vessels. For example, after exposure of corneas to silver/potassium nitrate, alkali, alloxan, colchicine, hyperthermal or radiofrequency burns, or intracorneal antigens in sensitized and nonsensitized animals, leukocytes enter the corneas before capillaries and accompany the angiogenesis. Leukocytes also precede blood vessels when rats are maintained on Vitamin A or riboflavin-deficient or high-tyrosine diets (152,263,287,369,417,418,813). These different models vary in several respects, including the degree of associated corneal edema and the latent period required for vascular invasion and the leukocytic infiltration manifests a variable time course after corneal injury. (ii) the location of the leukocytic invasion corresponds to the same site as the blood vessel invasion. (iii) the localization, depths of stromal involvement, and direction of the vascular invasion from the corneoscleral limbus correspond extremely well with the pattern of the leukocytic infiltration which these injuries induce. Discrete single corneal injuries such as silver nitrate or electrocauterization stimulate a localized leukocytic and vascular infiltration into the damaged corneal stroma from the nearest region of the corneoscleral limbus. Models that result in a diffuse, circumferential leukocytic infiltration into the cornea, such as topical alloxan or colchicine administration activate a vascular invasion into the periphery of the entire cornea. When leukocytes infiltrate the entire thickness of the corneal stroma, as when antigen is instilled into the corneas of sensitized animals, the vascular ingrowth extends throughout the depth of the corneal stroma. (iv) the amount of vascularization is also directly proportional to the leukocytic infiltrate. The injection of antigen into the corneas of sensitized animals, but not nonsensitized animals, provokes a marked leukocytic and vascular infiltration into the cornea to a degree which surpasses that of other models studied to date. (v) the onset of corneal vascularization is temporally related to the time of the initial leukocytic infiltration. For example, injection of antigen into sensitized animals produces an early leukocytic infiltration and corneal vascularization, while a longer latent period antecedes both the leukocytic and vascular invasion in rats on a riboflavin-deficient diet. (vi) the extent of the vascular ingrowth in the cornea is decreased in experimental situations in which the number of leukocytes in the corneal stroma is diminished (e.g., animals made leukopenic by whole-body x-irradiation, total lymphoid irradiation, antineutrophil serum, or ocular treatment with corticosteroids or various non-steroidal anti-inflammatory drugs (155,174,194,195, 264,694), or after the topical application of fatty acids that are structurally related to arachidonic acid (777). (vii) the intensity of new vessel formation within the cornea is directly related to the degree of leukocytic infiltration (63,203,263,264,275, 654,694). (viii) corneal vascularization is enhanced by conditions that promote leukocytes to infiltrate the corneal tissue (263). (ix) the angiogenic activity of intracorneal deposits of formyl methionyl leucyl phenylalanine appear to relate to the leukocyte attracting and activating properties of this compound (522). (x) splenic and bone marrow explants that are rich in leukocytes vascularize (744). (xi) following alkali burns it is noteworthy that when the burn affects the cornea, but not the corneoscleral limbus or conjunctiva, leucocytes and blood vessels invade the corneal stroma, but when the limbal conjunctiva is also burnt, less leukocytes infiltrate the cornea and almost no corneal neovascularization ensues (624).

It is also noteworthy as pointed out many years ago by Carrel (118,119), leukocytes and extracts of them contain substances capable of stimulating the proliferation of cells, such as fibroblasts. Evidence exists to implicate several different types of leukocytes in corneal angiogenesis and the relative importance of each cell type undoubtedly varies with the inducer of vascularization. New vessels can infiltrate corneas in an apparent absence of leukocytes and may not be essential for the initiation of corneal vascularization (194,694), but even then at least in some experimental models, leukocytes appear to facilitate or augment the process through incompletely understood mechanisms.

a.   Polymorphonuclear Leukocytes (neutrophil leukocytes, neutrophils)

Polymorphonuclear leukocytes, common companions of newly formed blood vessels in many tissues, including the cornea have been suspected of playing a role in angiogenesis (265,416,468,820). Not only does this cell type accompany the growth of blood vessels into the cornea in most experimental models that have been studied extensively to date, but they enter the tissue prior to angiogenesis (152,263). A heat labile fraction derived from polymorphonuclear leukocytes isolated from glycogen-induced peritoneal exudates, [which may have contained proteolytic activity], induced corneal neovascularization in 2 leukopenic weanling rats that died by the third day after receiving a massive dose of total body irradiation (1500 rads) (265). In a quantitative analysis of the infiltration of polymorphonuclear leukocytes into the rabbit cornea following thermal cautery, a direct relationship between the leukocytic infiltrate and the neovascular response has been noted (664). Peripheral corneal

31

burns elicit both a polymorphonuclear leukocytic infiltrate and neovascularization, whereas central burns evoke neither response (664). It is also noteworthy that corticosteroids, as well as those non-steroidal anti-inflammatory drugs (flurbiprofen and indomethacin) that suppress angiogenesis in the cornea also inhibit the polymorphonuclear leukocytic migration following corneal injury (128,718). The corneal and conjunctival polymorphonuclear leukocyte count is markedly reduced in rabbits with thermal cautery induced corneal neovascularization that is inhibited by prednisolone acetate (1%), indomethacin, or 0.3% flurbiprofen (636).

Factors that lead to the corneal infiltration of polymorphonuclear leukocytes varies with the experimental situation and local prostaglandin synthesis may be important following at least certain injuries (636) (Fig. 1).

The mechanism whereby polymorphonuclear leukocytes contribute to corneal angiogenesis, if they indeed do, remains to be established but several possibilities warrant consideration. Polymorphonuclear leukocytes, as well as other leukocytes and various other cells, are a source of cyclooxygenase and lipooxygenase products of arachidonic acid (88,351,719,722,790,799), including some with putative angiogenic activity such as PGE_1 (468). Also, polymorphonuclear leukocytes are a source of proteases including some that are implicated in angiogenesis, such as collagenase (468). A vast body of information now emphasises the importance of extracellular proteases in the regulation of cellular functions including the induction of mitogenesis in many cell types (665). Proteases may convert pro-growth factors to active moieties, act directly on cell surfaces causing transmembrane signals leading to stimulation of mitosis, and by inducing the secretion of other proteases in a multicascade of proteolytic events. Furthermore, extracts of polymorphonuclear leukocytes are weakly mitotic for umbilical vein endothelial cells (655).

Despite the aforementioned, other investigators have been unable to induce neovascularization in the cornea by injecting peritoneal polymorphonuclear leukocytes into the rabbit (139,550) or guinea pig cornea (139,607). However, even if polymorphonuclear leukocytes are not essential for the initiation of corneal vascularization (194,694) the current consensus is that they may play a facilitatory role.

Some studies that claim to induce corneal vascularization in the absence of leukocytes (194,694) did not exclude the presence of polymorphonuclear leukocytes in the tears where they are known to reside follow any corneal injury. In studies aimed at evaluating the contribution of polymorphonuclear leukocytes in corneal vascularization Hoban and Collin (354) produced a thermal burn in the center of the cornea of rats and reduced the emigration of neutrophils from the limbal blood vessels with topically administered sodium salicylate and prednisolone. They found that the sodium salicylate drops increased the length of invading blood vessels compared to non-treated cauterized corneas, but noted no consistent relation between the number of extravascular neutrophils at the corneoscleral limbus and the extent of corneal blood vessel growth. However, in this study the leukocytic infiltrate was only quantitated in small pieces of the peripheral cornea and the mean lengths of the invading corneal blood vessels were determined in the experimental groups, rather than total areas of neovascularization.

Even if polymorphonuclear leukocytes contribute to corneal angiogenesis in someway this cell type is frequently present within the pathologic cornea in the absence of associated new blood vessels.

b.   Lymphocytes

An association between vascular proliferation and lymphoid cells has been recognized for decades in non-malignant situations.   For example, vascular proliferation is a feature of some skin lesions, such as cutaneous lymphoplasia (lymphocytoma) (485) and plasmacytosis (582).   Angiogenesis is also a prominent feature of a condition characterized by a marked proliferation of capillaries and the entire range of immunologically reactive cells (including lymphocytes, plasma cells, and immunoblasts [large transformed lymphocytes]) within lymph nodes [angioimmunoblastic (immunoblastic) lymphadenopathy] (480,619,675).   Despite these associations of angiogenesis with lymphoid cells, the possibility that lymphocytes might induce new vessel formation did not receive attention until relatively recent times.   Much evidence implicates lymphocytes in angiogenesis in tissues other than the cornea and a vast body of evidence supports the concept that activated lymphocyte can induce corneal neovascularization (23,26,27,55,56,203,204,394,482, 556,563,696,697,793).   Sidky and Auerbach (697) have documented that the intradermal transfer of immunocompetent lymphocytes induces endothelial cell proliferation.   The intrastromal implantation of homologous, but not autologous, adult lymph nodes (238) and allogeneic lymphocytes (203) induce corneal angiogenesis in the rabbit.   Corneal neovascularization has also been induced in the mouse with corneal implants of whole lymph node fragments (563) and concanavalin A stimulated allogeneic lymphocytes (204).   Using inbred mice Epstein and Stulting (158) have provided evidence that strain-related differences in reactivity and host recognition of histocompatibility differences are both important in this response.   Aside from stimulating vascular endothelial proliferation, lipopolysaccharide as well as concanavalin A (con A) stimulated leukocytes stimulate both DNA and protein synthesis by corneal fibroblasts (keratocytes) (57).

Not all investigators have been impressed with the ability of lymphocytes to induce angiogenesis.   Fromer and Klintworth (265) found that *non-activated* lymphocytes isolated from thymus, spleen, and lymph nodes of normal rats did not elicit corneal vascularization in weanling rats with severe radiation induced leukopenia.   Without providing experimental data Polverini, Cotran, Gimbrone and Unanue (607) reported that they were unable to produce corneal neovascularization by implanting activated lymphocytes into guinea pig corneas.   In a later publication from this laboratory Gimbrone and colleagues (286) reported no neovascularization in 58 corneas following the injection of normal and immunologically primed lymphocytes immediately after isolation with and without the purified protein derivative of tuberculin (PPD).

How lymphocytes induce angiogenesis remains to be determined, but following stimulation by mitogens they undergo blastogenesis and secrete a variety of biologically active substances, including some that have been implicated in angiogenesis: prostaglandins (55,250) and lymphokines (23).   Activated T lymphocytes may mediate angiogenesis by activating resident macrophages (see below) and by the production of a proliferation-inducing endothelial cell lymphokine (ECL-1) (793). Interferon γ produced by activated T cells may also contribute to the induction of angiogenesis by increasing the release of superoxide by endothelial cells (502).

c.   Macrophages

Several observations suggest that activated macrophages play a role in angiogenesis (42,267,286,356,396,423,426,427,455,456,501,556,587,607,609,750,788, 812).   (i) Endothelial proliferation occurs at the height of the delayed sensitivity reaction in the skin of guinea pigs (608). (ii) Macrophages may be the only inflammatory cells at a corneal wound site capable of initiating new vessel formation (550). (iii) Intracorneal injections of activated macrophages, or conditioned medium derived from them in culture, produce corneal vascularization in a high percentage of animals (60 to 80%) (607). (iv) Macrophages can be stimulated to produce a potent growth factor (macrophage derived growth factor) that induces neovascularization when implanted into the cornea (286) and which stimulates the proliferation of vascular endothelium, fibroblasts and vascular smooth muscle in culture (501). (v) Using the rabbit corneal implant lactate, but not pyruvate, has been found to cause macrophages to secrete angiogenic factors (385). The angiogenic activity of macrophage-conditioned media is enhanced following exposure of cultured macrophages to bacterial endotoxin (*Escherichia coli* lipopolysaccharide), latex particles, viable microorganisms (*Listeria monocytogenes*) or phorbol myristate acetate (286).

Macrophages are not essential for new vessel growth in nonimmunologic acute inflammation (681). Not all populations of monocytes appear to induce neovascularization in the cornea. Monocytes obtained from the citrated blood buffy coat of healthy adult human donors fails to stimulate angiogenesis in the rat cornea, whereas monocytes activated with concanavalin A (25 µg/ml) or endotoxin (5 µg/ml) for 20 hours are potently angiogenic (426). Antigenically stimulated macrophages secrete increased amounts of mitogens (293), but hypoxic macrophages do not (385).

The question of how macrophages induce angiogenesis remains to be established, but current evidence points to several possibilities. Activated macrophages produce several cytokines that have been implicated in angiogenesis (Fig. 2). These include interleukin 1 (542), fibroblast growth factor (317), tumor necrosis factor α (258,456,602), tumor necrosis factor β (635), monocyte/macrophage derived growth factor (286,455,501), angiotropin (356), and human angiogenic factor (HAF) (a 67 kDa protein with a pI of 5.0 that does not have a marked affinity to heparin) (267). In addition to cytokines, macrophages release proteases (566,622), and as discussed elsewhere extracellular proteases may contribute to angiogenesis because of their capability to induce mitogenesis and degrade extracellular matrix components.

 d. Platelets

During the early hours of the inflammatory response, as after chemical injuries to the cornea, platelets are prominent in the dilated stagnant blood vessels (438). As a normal response to injury these cellular elements aggregate and degranulate releasing numerous inflammatory mediators from the alpha-1 granules and presumably by other mechanisms as well. Platelets release substances that stimulate DNA synthesis and cell migration in vascular endothelium (141,259,408,484,788,789) (Figs. 3-5). These biologically active substances, which may play a role in corneal angiogenesis, include platelet derived growth factor (PDGF) (167,826), TGF-α (643), TGF-β (22,634,643), adenosine triphosphate (ATP) (256) 5-hydroxytryptamine (256) and platelet derived endothelial-growth factor (PD-ECGF, ENDO-GF) (378,408). Adenosine triphosphate (ATP) and 5-hydroxytryptamine (serotonin) both induce cell proliferation of cultured human umbilical vein endothelium and elicit a positive angiogenic response on the chick chorioallantoic membrane (256). PD-ECGF, which differs from PDGF, is also mitogenic for vascular endothelial cells (378,408). Aside from containing putative angiogenic factors platelet α-granules contain a protein (platelet factor-4) that inhibits angiogenesis in the chicken CAM system (495). This inhibition of angiogenesis is abrogated by heparin - a glycosaminoglycan for which it has a high affinity.

e. Mast Cells

Mast cells, which like many other cell types are now known to manifest significant phenotypic heterogeneity (269), are best seen in histologic sections with the aid of special stains (such as toluidine blue, the Giemsa stain and the recent immunoperoxidase technique that employs murine monoclonal antibodies against specific neutral proteases) (160). For a long time the mast cell has been suspected of having more than a passive role in new vessel formation (648,649), and this cell type has frequently been identified in association with angiogenesis of the cornea (707,709,831), of tumors (256,400) and of other tissues (498). Others have also suggested that mast cells may influence the growth of vascular endothelium (648). The mast cell secretagogue compound 48/80 stimulates angiogenesis in a dose-dependent manner in a mesenteric window assay (577). Kessler, Langer, Pless and Folkman (406) observed that mast cells accumulate markedly before the ingrowth of new capillaries at tumor sites and in the chick CAM under circumstances which elicit a strong vasoproliferative response. Also, mast cell granules stimulate the proliferation of microvascular endothelial cells in culture (498). Rat mast cells also evoke an angiogenic response when implanted on the chick CAM (406).

Because mast cells are present at the corneoscleral limbus, the possibility of them inhibiting corneal vascularization was raised several years ago by Smith (707), who proposed that their massive destruction following corneal injury caused blood vessels to invade the cornea. Mast cells infiltrate the cornea before blood vessels and have been implicated in capillary growth as part of the host response to bovine intralamellar heterografts in the rabbit (709). However, they are not a prominent feature in most cases of corneal vascularization (276) and since angiogenesis can be elicited in mast cell deficient WW^V mutant mice (24), mast cells are clearly not essential for angiogenesis.

Recent studies have particularly implicated two products of mast cells in angiogenesis (Figs. 5 and 6): heparin and histamine (498). Heparin and several heparin fragments promote angiogenesis in vivo (244). Heparin elicits an angiogenic response on the chick CAM according to some investigators (255,521,522), but heparin alone does not induce angiogenesis in the CAM in the experience of others (246). Heparin also elicits neovascularization in the rabbit cornea (521,522) and augments angiogenesis induced by tumor (406) in the CAM assay. This effect occurs with the anticoagulant and non-anticoagulant versions of heparin indicating that it is independent of the coagulation cascade (122). Heparin that is incorporated in Elvax pellets (100 µg/ml) induces new vessel growth in the cornea according to some investigators (522), but others find heparin only to be angiogenic in the cornea when complexed to copper (8,617,618). A heparin:copper complex is also chemotactic for bovine capillary endothelium in vitro (8). Heparin increases the production of plasminogen activator, a protease implicated in angiogenesis, by adipocyte-induced stimulated vascular endothelial cells in culture (122). Heparin also promotes the migration of capillary endothelial cells (8,34,232,522), but not aortal endothelial cells (521,522). At low concentrations of heparin (20 µg/ml) McAuslan, Reilly, Hannan and Gole (522) found no migration activity towards either capillary or aortal endothelial cell lines, but at high levels of heparin (100 µg/ml) they found a consistently small, but positive effect on only capillary endothelial cells. This activity is blocked by protamine sulfate, a protein that binds avidly to heparin (34,748). Protamine also inhibits angiogenesis associated with embryogenesis, inflammation and some immune reactions (748). While heparin does not promote endothelial cell proliferation in some systems (34,122,748), it enhances the growth of some strains of vascular endothelial cells derived from the vena cava and tibial artery. Several fragments of heparin stimulate DNA synthesis in cultured endothelial cells derived from the cerebral microvasculature (51). Heparin also potentiates the proliferative effect of acidic fibroblast growth factor on endothelial cells in culture (760). Heparin also protects some growth factors, such as fibroblast growth factor, from inactivation (308). Relevant to the potential role of mast cells in angiogenesis is the avid binding of heparin to a family of growth factors with putative angiogenic activity in a variety of assay systems (166,235,241,412,490) and this discovery made heparin-affinity chromatography useful in the purification of certain angiogenic growth factors (687). Heparin enhances the binding of endothelial cell growth factors to surface receptors on these vascular cells (672). How this occurs still needs to be established, but heparin inhibits fibronectin synthesis by endothelial cells (483).

The putative role of heparin in angiogenesis is clearly complex. Despite its potentiating effect on new vessel formation, heparin, and several non-anticoagulant fragments of heparin, are reported to inhibit the angiogenesis in several systems (51,161,244,247,375). This includes new vessel growth induced by tumor in the CAM and cornea in the presence of cortisone or hydrocortisone (244), or on the CAM by other corticosteroids (161,375). Heparin, as well as heparin in combination with hydrocortisone has a slight inhibitory effect on DNA synthesis and cell growth of mouse cerebral endothelial cells (51).

Since the secretory granules of all populations of mast cells studied to date contain serine proteases (682) mast cells may also contribute to the degradation of basement membranes during angiogenesis. At least one of these proteases, chymase, is ionically bound to negatively charged heparin proteoglycans when both are exocytosed from immunologically activated mast cells, and the activity of the enzyme is substantially inhibited against large substrates when it is bound to heparin proteoglycan (682).

As discussed elsewhere in this review, corneal edema is associated with corneal vascularization and mast cells contain biogenic amines, such as histamine and serotonin, which can contribute to this edema.

2.   Humoral Aspects of Inflammation

In addition to the cellular infiltration that characterizes the inflammatory response many biologically active substances are liberated within the injured corneal tissue including eicosanoids, biogenic amines and plasminogen activator. Other substances are released from cells that enter the cornea (interleukin 1, interleukin 2, macrophage derived growth factor, angiotropin, tumor necrosis factor α, interferon and other cytokines). Also, as a consequence of the increased permeability of the pericorneal blood vessels during acute inflammation, many constituents of plasma gain access to the cornea. These include fibrinogen, plasma fibronectin and a wide variety of growth factors (including mesodermal growth factor, insulin, platelet derived growth factor, epidermal growth factor and fibroblast growth factors) capable of enhancing the proliferation of various cell types, including vascular endothelium.

a. Biogenic Amines (Histamine, Bradykinin and Acetylcholine)

Histamine, a potent component of the inflammatory response, and a constituent of mast cell granules, was the suspected angiogenic factor of Campbell and Michaelson (116). Zauberman, Michaelson, Bergman and Maurice (822) obtained evidence to support this view when they placed a plastic tube into the cornea of rabbits and found that various biogenic amines (acetylcholine, histamine, serotonin or bradykinin) caused blood vessels from the corneoscleral limbus to grow into the lumen of a tube in one third of the rabbits into whose corneas these biologically active substances were perfused. That histamine may contribute to angiogenesis also finds support in the observation that this biogenic amine is mitogenic to human microvascular (498) and umbilical vein endothelial cells in culture (256). Studies with specific agonists (2-pyridylethylamine dihydrochloride, Dimaprit) and antagonists (Metiamide Clemastine fumarate, mepyramine maleate) of histamine suggest that this effect is mediated through an H_1 receptor (498). Histamine may also conceivably enhance angiogenesis through the resultant increased vascular permeability that it causes in the inflammatory response, and mast cells which accompany angiogenesis may be a source of this histamine.

Nunziata, Smith and Weimar (579), however, could not produce or stimulate the vascular labelling of the pericorneal microvasculature with carbon in rats with or without a radiofrequency burn of the cornea after topical bradykinin or histamine.

b. Plasminogen Activator (Urokinase) and other Proteases

Because capillary endothelial cells penetrate through the basal lamina of the blood vessels from which they arise and invade the surrounding connective tissue proteases, such as plasminogen activator and collagenase, are suspected of contributing to new vessel formation. This concept finds support in several observations. Certain stimulators of angiogenesis, including phorbol esters and basic fibroblast growth factor, induce the synthesis and secretion of proteases by vascular endothelial cells (322,323,547,553). Polymorphonuclear leukocytes and other cells that have been incriminated in corneal angiogenesis are also a source of proteases, such as collagenase (468). The ubiquitous plasminogen activator, a serine protease crucial to the fibrinolysis of thrombi, has been implicated in localized extracellular proteolysis, and in corneal neovascularization (63,297). Blood vessels grow into the cornea following the intracorneal instillation of plasminogen activator and, as with intracorneal injections of other substances, this reaction, is accompanied by a leukocytic infiltration. Moreover, the reaction is diminished if the urokinase is inactivated by heat or by an inhibitor of a specific active site of plasminogen activator (Phe-ala-arg-chloro-methylketone) (63). Although it is conceivable that the angiogenic activity of plasminogen activator reflects proteolysis, this enzyme is also known to contain amino acid sequences that are similar to those in portions of epidermal growth factor. Capillary endothelial cells (bovine) secrete both tissue type and urokinase type plasminogen activator in culture and the secretion of both of these enzymes is enhanced by the addition of a heparin binding angiogenic factor (basic fibroblast growth factor) (821). It is noteworthy that the corneal epithelium contains significant amounts of tissue plasminogen activator (62,278,449,591,791) and so does the corneal stroma (278) and endothelium (216,278,449,591).

The potential role of proteases in angiogenesis is obviously complex. In the presence of plasminogen, or other fibrinolytic proteases, vascular endothelial cells may not invade a fibrin matrix (546). The protease thrombin potentiates the growth response of human vascular endothelial cells in culture to fibroblast growth factor (307).

c. Eicosanoids (Prostaglandins, and other Metabolites of Arachidonic Acid)

Soon after tissue injury, prior to the onset of an obvious inflammatory response, phospholipids within cells become degraded by phospholipases to liberate nonesterified fatty acids, including arachidonic acid (Fig. 7). The latter is metabolized further by the cyclooxygenase pathway to prostaglandins (PG), thromboxanes, and prostacyclin and by lipooxygenase pathways to produce hydroxyeicosatetraenoic acids (HETEs) and leukotrienes. Angiogenic activity has been attributed to prostaglandins and notably prostaglandin E_1 (PGE_1) (55,468,617,829). That prostaglandins may participate in the cascade of events leading to corneal vascularization finds support in several observations: (i) BenEzra (55) observed that the intracorneal instillation within Elvax pellets of several prostaglandins induced angiogenesis (PGE_1 > PGE_2 and $PGF_{2\alpha}$ inconsistent). (ii) Ziche and colleagues (828) found that a pellet bearing 1 µg of PGE_1 induces corneal vascularization when implanted into the rabbit cornea, but that PGE_2 was not angiogenic in this system at the same dosage. The angiogenic effect of intracorneal PGE_1 has been found to be enhanced by the addition of mono- and trisialotetraesosyl ganglioside sodium salts (827). (iii) An elevated PG concentration has been noted in the aqueous humor during corneal angiogenesis secondary to experimentally produced anterior uveal ischemia (539). (iv) Corneal neovascularization is suppressed by inhibitors of arachidonic acid metabolism that block PG synthesis. Such inhibition follows treatment with corticosteroids (20,128,155,183,264,370,416,515,533,572, 586,677), cyclooxygenase inhibitors (155,174,183,266,333,337,494,539,593,636,700), and fatty acids that are structurally related and arachidonic acid (777).

Alessandri, Raju and Gullino (8) provided evidence that the angiogenic effect of PGs (especially PGE_1 and less conspicuously PGE_2) may involve an interaction with corneal tissue. They inserted a PGE_1 containing slow release copolymer levix into a rabbit corneal pocket. After little more than 3 days pieces of the cornea from the region between the implant and the pericorneal vessels were removed. These were then placed in wells and dishes containing gelatin-agarose at a distance from other wells containing capillary endothelium to test for migrating properties or put in a Boyden chamber with the same objective. Corneas exposed to PGE_1 in this way induced endothelial cell migration whereas PGE_1 alone and cornea exposed to PGE_2 had much less of an effect.

Most cells studied to date (819), including those of the cornea and other ocular tissues (49,50,75,77,82,400), synthesize PGs. These cyclooxygenase products can form within the cornea in the absence of exogenous cells, but additional potential sources of their production exist in situations in which corneal angiogenesis takes place. These include activated lymphocytes (55), macrophages (371) and aggregated platelets (688), which are known to accumulate in the pericorneal vasculature prior to angiogenesis in some corneal injuries (524). The high content of leukocyte phospholipase provides inflammed tissue with a possible source of PGs.

Even if PGs contribute to corneal angiogenesis this effect may be mediated by leukocytes, since those cellular elements not only synthesize PGs and chemotactic leukotrienes (49,50), but polymorphonuclear leukocytes are attracted chemotactically to metabolites of arachidonic acid (3,350,435). Also, corneas that are markedly vascularized following implants sequestering PGE_1 are infiltrated by polymorphonuclear leukocytes (55). Drugs such as corticosteroids and non-steroidal anti-inflammatory drugs that block PG synthesis suppress the leukocytic infiltrate as one would expect because of the chemotactic effect of certain PGs on polymorphonuclear leukocytes (3,350,435).

Another potential role of PGs in angiogenesis involves their ability to mobilize copper ions, which have been implicated in corneal neovascularization (8,515,617). Indeed some investigators consider copper ions necessary to induce new vessel formation in vivo and endothelial cell migration in vitro (519,618). Prior to the onset of corneal angiogenesis following the implantation of intracorneal pellets containing PGE_1 the corneal copper level increases and the tissue synthesizes several enzymes including the copper-dependent benzylamine oxidase (828). Moreover, a dietary deficiency of copper is stated to prevent PGE_1 induced corneal angiogenesis (829). Several small copper-containing molecules trigger angiogenesis and this property is lost when the copper is removed (616,618). How copper is involved in angiogenesis remains unclear, but copper ions stimulate collagenase production by the polymorphonuclear leukocyte in a dose dependent manner and this enzyme has been implicated in angiogenesis (468).

Leukotrienes and other metabolites of arachidonic acid may also play a role in angiogenesis. Leukotriene D4 and HETE (hydroxyeicosatetraenoic acid) induce endothelial cell migration (521) and leukotriene C4 promotes bovine aortic endothelial cell growth in culture (541). Polymorphonuclear leukocytes possess specific membrane receptors for leukotriene B4 which is a potent chemoattractant for this cell type (76) and it is noteworthy that the implantation of pellets with leukotriene B4 in the rabbit induce a marked leukocytic infiltration but without neovascularization (637). 12-HETE is also chemokinetic and chemotactic for polymorphonuclear leukocytes (76).

      d.   Fibrin

During the increased pericorneal vascular permeability that accompanies the complex web of reactions of inflammation fibrinogen enters the extracellular space of the cornea along with fibronectin and other plasma constituents (Fig. 8). Fibrinogen is a heterodimer composed of paired Aα, Bβ and γ chains (555) and following extravasation the A and B fibrinopeptides are rapidly cleaved to form fibrin monomeres which polymerize spontaneously to form cross-linked fibrin as tissue procoagulants interact with plasma clotting factors (188). In many tissues, fibrin has been implicated as a proinflammatory mediator following its activation by other systems, such as complement, and as a primary trigger of such a reaction.

Factors regulating fibrin dissolution in the cornea have not been investigated, but plasminogen activator and other proteases secreted by inflammatory and other corneal cells may play an important role. For several reasons fibrin and products of its degradation are suspected of being important in angiogenesis. Fibrin appears within the inflammatory exudate of vascularizing corneal tissue (524) and the invasion of corneal tissue by blood vessels is apparently enhanced by antifibrinolysins (epsilon-aminocaproic acid and BP 961) (708). Moreover, fibrin deposition consistently precedes new vessel growth in solid tumors and healing wounds (188) and fibrin degradation products induce angiogenesis in the chick chorioallantoic membrane (759). A fibrin gel around tumors progressively vascularizes (185-187) and subcutaneously implanted fibrin gels induce an angiogenic response in the absence of tumor cells and platelets (188). Vascular endothelial cells form a confluent monolayer on fibrin gels, but invade it following treatment with 4β-phorbol 12-myristate 13-acetate (PMA) in the absence of plasminogen or fibrinolytic inhibitors

(546). Fibrin gels placed subcutaneously in animals are invaded by fibroblasts and blood vessels which gradually replace the gels (186). Other lines of evidence support a role for the fibrinolytic system in angiogenesis (821). The addition of plasminogen to a type 1 collagen gel matrix increases the length of bovine capillary endothelial cell tubes in culture and inhibitors of plasmin suppress the enhancing effect of plasminogen (821).

e. Cytokines

Several well-defined proteins that are synthesized by lymphocytes, mononuclear phagocytes and other cells mediate some aspects of the inflammatory response (604). Some of these cytokines have been implicated in angiogenesis and cause vascular endothelial cells to synthesize other factors (Fig. 9).

Interleukin 1: Information about interleukin 1 (IL-1), which was originally designated lymphocyte-activating factor (282) has been extensively reviewed (176,177,452,588,589). This potent soluble intercellular mediator of inflammatory and immunological reactions, is elaborated in response to appropriate stimuli by the corneal epithelium, monocytes/macrophages, vascular endothelial cells, and virtually every nucleated cell type (176,453,723,784,808). Molecular cloning studies have established the existence of at least two species of IL-1 (IL-1α and IL-1β) and both forms of the cytokine lack signal peptide sequences that are characteristic for secretory proteins (453). While initially defined as a stimulator and activator of T lymphocytes IL-1 has numerous biological effects including the activation of endothelial cells to express surface proteins (72,271,603) that increase the adhesiveness of the vascular endothelium for peripheral blood polymorphonuclear leukocytes and monocytes (72,73) (Fig. 10). Aside from producing IL-1, vascular endothelial cells also respond to IL-1. IL-1 also induces cultured human umbilical vein endothelial cells to produce a tissue factor-like procoagulant activity (70-73) and fibroblasts to proliferate (441,558). With regard to its growth promoting activity it is noteworthy that IL-1 has approximately 30 percent amino acid sequence similarity with two endothelial cell mitogens: α fibroblast growth factor 1 (382) and endothelial cell growth factor (382). IL-1 is an extremely potent chemotactic agent of polymorphonuclear leukocytes (163,453) and it stimulates the metabolic activity of these cells (453). The intracorneal introduction of IL-1 induces corneal angiogenesis (59,178,340) and at least part of this angiogenic stimulation by IL-1 may be secondary to an effect on polymorphonuclear leukocytes, especially since IL-1 inhibits endothelial cell proliferation (486).

Interleukin 2: Interleukin 2 (IL-2), while producing vascularization after its introduction into the rabbit (330) or mouse (202) cornea, appears to be less potent than interleukin 1 (59,340).

Interleukin 6: The 26-kDa cytokine interleukin 6 (IL-6) that is produced by and acts on different cell types, including monocytes, T lymphocytes and vascular endothelium. Although the potential role of this cytokine in corneal neovascularization has not been evaluated IL-6 mRNA is produced *in vivo* during the physiologic angiogenesis that accompanies ovarian follicle formation and the maternal decidua following embryonic implantation suggesting a role for IL-6 in angiogenesis (558).

Monocyte/Macrophage Derived Growth Factor (MDGF): MDGF stimulates the growth of certain non-lymphoid mesenchymal cells, including fibroblasts and smooth

40

muscle cells and in certain instances vascular endothelium in culture (500,501). It is structurally related to, if not identical to, basic fibroblast growth factor (251).

Angiotropin: Activated macrophages produce a copper-containing polyribonucleopolypeptide (~ 4.5kDa) (angiotropin), which enhances the random migration, but not the proliferation, of capillary endothelial cells. In the presence of angiotropin capillary endothelial cells, but not aortic endothelial cells, form tube-like structures on gelatinized plates (356).

Tumor Necrosis Factor α (TNFα): TNF, which was first detected in the serum of Bacillus Calmette-Guerin (BCG) primed endotoxin-treated animals, is a nonglycosylated polypeptide with a molecular weight of approximately 17 kilodaltons. Several reviews have been written on this cytokine (66,453,585). This product of macrophages lacks the typical 20-30 amino acid long hydrophobic signal peptide sequence characteristic of secretory proteins (453). TNFα induces inflammation and a corneal angiogenic response when introduced into the cornea (258,456) or when injected intravitreally (641). TNF also induces a neovascular response on the CAM (456) and causes cultured bovine adrenal capillary endothelial cells to form capillary-tube like structures on collagen gels (456). The angiogenesis induced by TNFα is probably mediated indirectly because this cytokine does not affect endothelial cell proliferation *in vitro*. TNF inhibits cultured capillary (678), aortic arch (258) and other (725) endothelial cell proliferation and this cytokine is cytotoxic against bovine capillary endothelial cells (661,662). TNF also inhibits the growth of capillaries from epididymal fat pads seeded onto a confluent culture of myofibroblastic cells (660) and this inhibition is blocked by a monoclonal antibody against TNF (660). Moreover, TNF triggers IL-1 production by endothelial cells (467,567). TNF is also chemotactic for neutrophils and stimulates their activation as measured by the respiratory burst and degranulation (414,415,770). Like IL-1, TNF activates vascular endothelial cells to express neutrophil adhesive proteins (271,603).

Interferon: Human leukocyte α interferon inhibits the migration of bovine endothelial cells (98). The lymphokine immune γ interferon induces cultured human umbilical vein endothelial cells to express class II histocompatibility antigens (Ia antigens) (605,606). Although the effect of interferon has not been fully evaluated in the cornea it may be important in allograft rejection, because Ia-bearing cells are much more immunogenic than Ia negative cells (798). Passenger leukocytes within corneal vessels may contribute to the increased risk of vascularized corneas to graft rejection. Currently there is no evidence that the interferons are angiogenic, however, interferon γ inhibits capillary growth from rat epididymal fat pads *in vitro* (772).

B.   Corneal Constituents

The possibility of an angiogenic mediator arising from injured corneal tissue has been raised and this stems from the old concept of injured cells producing growth factors. Almost a century ago Wiesner postulated that injured cells might liberate substances capable of stimulating cell proliferation. Much later Abercrombie (1) suggested that a "wound hormone" released locally by damage to cells in response to local injury might stimulate mitotic activity of the tissue around the wound. If corneal cells produce an angiogenic factor they clearly do not liberate significant quantities of it under normal circumstances. The epithelium constitutes the major cellular constitute of the cornea and it is hence the prime suspect from the standpoint of this tissue. However, because blood vessels can grow into corneas devoid of an epithelium (277,823), the epithelium is apparently not essential for corneal neovascularization. Several observations suggest that a reactive corneal epithelium may contribute to an angiogenic response: (i) a corneal epithelial injury frequently coexists with an underlying stromal neovascularization; (ii) heat stable homogenates of freshly excised or cultured normal corneal epithelial cells induce angiogenesis when infused into the peripheral cornea of rabbits (194,195,197); (iii) corneal epithelial cells and medium from cultures of them (containing a heat stable mitogen) stimulate the growth of cultured vascular endothelium (derived from rabbit marginal ear veins, but not rabbit aorta) in a dose dependent manner (195,196); (iv) the vascular endothelium in thermally injured skin has a higher labeling index (^{3}H-thymidine) beneath foci of incomplete reepithelialization compared to nonepithelialized injured sites (689); (v) a heat-labile, non-dialyzable apparently polar extract from skin epidermis incites angiogenesis in the hamster cheek pouch (814).

The mechanism for the putative angiogenic role of the corneal epithelium remains unknown and may involve factors considered elsewhere in this review. For example, the corneal epithelium is capable of synthesing prostaglandins and plasminogen activator (62,127,278,449,591). Also, the corneal epithelium normally contains vast amounts of acetylcholine, which stimulates the formation of phosphatidic acid, an obligatory intermediate in the inositol phospolipid pathway (613). Inflammatory mechanisms may also account for some observations that implicate the corneal epithelium in angiogenesis. Corneas perfused with an epithelial homogenate also induce a dense leukocytic infiltrate (mononuclear and polymorphonuclear leukocytes) within the corneal stroma (195).

Relevant to the possibility that corneal fibroblasts produce an angiogenic factor is the observation that medium conditioned by fibroblasts is mitogenic to vascular endothelial cells (398).

In experiments in which open plastic tubes were implanted radially into rabbit corneas, Maurice, Zauberman and Michaelson (512) observed that by repeatedly wounding the tissue close to the central end of the open, but not closed, tubes, vessels grew into the open plastic tubes. This was interpreted as evidence in favor of a vasostimulatory factor being liberated at the wound and passing down the lumen of the tube to the responding vessel walls. Maurice, Zauberman and Michaelson (512) suggested that cells produced a vasostimulatory factor and that this diffuses in direct proportion to the molecular weight of the particles according to the laws of diffusion. If an angiogenic factor arises in damaged tissue from local constituents, one would anticipate that a particular injury would consistently provoke corneal vascularization, but similarly treated corneal tissue that is implanted into hamster cheek pouches does not evoke angiogenesis reproducibly (416).

C.   Cells of the Microvasculature

A potential source of the factor that initiates corneal angiogenesis that has not received as much attention as it deserves is the endothelium of the pericorneal microvasculature from which the new vessels arise. A wide variety of responses take place in the endothelium of the microvasculature during inflammation. Following activation by a wide variety of inflammatory mediators the vascular endothelium produces many specific biologically active proteins. These products include: tissue factor, fibronectin (517), von Willebrand factor, and inhibitors of fibrinolysis (650). Endothelial-leukocyte adhesion proteins [Endothelial-leukocyte adhesion molecule 1 (ELAM-1), intercellular adhesion molecule 1 (ICAM-1) and 2 (ICAM-2) and vascular cell adhesion molecule 1 (VCAM-1)] are involved in causing the adherence of polymorphonuclear leukocytes, lymphocytes or monocytes to the vascular endothelium (74,604) (Fig. 11). IL-1β and TNF induce human vascular endothelial cells to synthesize and secrete a monocyte chemoattractant (MCP-1/JE) which may contribute to the accumulation of monocytes at sites of inflammation (639). Aside from mast cells the vascular endothelium itself produces a biologically active heparin-like species (121). The changes in the endothelial cells include active participation in the increased microvascular permeability (651,652) and their loss following cytolysis (761). During the inflammatory response the vascular endothelium can be injured passively by microorganisms or leukocytes or actively to specific mediators such as bradykinin, cytokines (including interleukin 1) or thrombin (650). Plasminogen binds to cultured human umbilical vein endothelial cells and in this immobilized state becomes a substrate for efficient conversion to plasmin by tissue plasminogen activator (328).

The endothelium of the microvasculature may contribute directly to the digestion of the basal lamina that normally surrounds it as well as of the adjacent extracellular matrix during the endothelial migratory aspect of angiogenesis (Fig. 12). This notion is supported by several observations of proteolytic activity by vascular endothelium. Endothelial cells release the two major forms of plasminogen activator, tissue plasminogen activator and urokinase (328,464). Activated rabbit brain capillary endothelial cells have also been shown to synthesize metalloproteinases (procollagenase and prostnomelysin) (347). Bovine aortic endothelial cells secrete a platelet derived growth factor-like protein growth factor (175). Fibroblast growth factor stimulates bovine adrenal gland capillary endothelial cells to produce a urokinase-type plasminogen activator, a protease implicated in angiogenesis (547). Vascular endothelial cells migrating from the edges of a wounded confluent monolayer of cells in culture produce a quantitative increase in plasminogen activator (595). Basic fibroblast growth factor (bFGF), which is abundant in a wide variety of highly vascularized tissues and tumors, is not only mitogenic for capillary endothelial cells, but such endothelial cells also produce bFGF in culture themselves and express the bFGF gene. Because of this it is conceivable that capillary endothelial cells induce new capillaries by a self-stimulatory (autocrine) fashion (679). Even sprouting endothelial cells may produce factors that modulate themselves in an autocrine fashion.

D.   Extracellular Matrix

Studies on angiogenesis in several situations suggest that the extracellular matrix may play a cardinal role in the formation of new blood vessels and orchestrate the growth and differentiation of vascular endothelial cells (237,251,376,491,493,545, 669,670). The extracellular matrix influences gene expression, by incompletely understood mechanisms, but which may involve the binding of matrix constituents to specific receptors located at the cell surface (729). A functional relationship between the vascular endothelium, extracellular matrix and certain growth factors seems to exist. For example, capillary endothelial cells will proliferate in response to epidermal growth factor only if they are plated on fibronectin (514) and to other angiogenic factors if plated first on native collagen (404,522). Several studies have shown that collagenous components of the matrix influence the ability of capillary and large vessel endothelial cells to form capillary-like tubular structures (493,543-545,547). The proliferative capability and biosynthetic activity of microvascular endothelial cells in culture is also influenced by components of the extracellular matrix (376,669,670). Form, Pratt and Madri (251) found that the proliferation of vascular endothelial cells over a 5 day period was significantly greater on laminin than on either plasma fibronectin, interstitial collagen types I and III, or on basement membrane collagen type IV.

Fibronectin, a product of fibroblasts, is not only synthesized by the cornea, but can enter this tissue from the plasma (Fig. 13). Fibronectin or fragments of it, are chemoattractic for endothelial cells (89), and blood monocytes (578). Some cells, such as monocytes, bear membrane receptors for surface-bound fibronectin (69) and these may serve to stimulate macrophages to generate products such as macrophage derived growth factor (500), plasminogen activator (68) and elastase (68). Fibronectin enhances capillary endothelial cell migration in cell culture and heparin enhances this effect (774). Moreover, fibronectin antiserum blocks the migration of the endothelial cells and neutralization of the antiserum restores the mobilization (774).

Although evidence to implicate specific bacterial products in corneal angiogenesis is not yet available, a bacterial product of *Bartonella bacilliformis* has been shown to stimulate the proliferation and increase the production of tissue plasminogen activity in human vascular endothelial cells *in vitro* and to stimulate the formation of new blood vessels in a subcutaneous wound healing assay (272).

E.   Non-Corneal Cells

It seems unlikely that a mediator of corneal vascularization would arise from the tears or aqueous humor since it would probably be considerably diluted by the time it reached the cornea. Moreover, such a source should not promote focal areas of corneal angiogenesis. Nevertheless, Szeghy (738) produced corneal vascularization by the intracorneal injection of lacrimal fluid from rabbits and humans and maintained that prior boiling eliminated the effect. In noncorneal tissue a variety of cell types other than leukocytes are suspected of producing angiogenic factors. These include neoplastic cells (7,93,222,223,241,245,283,405,411,436,438,599,668,686, 687,795), the retina (37,45,131,164-166,215,288,324,409), platelets (141,167), the retinal pigment epithelium (481) and the skin (epidermis) (581,649,814).

F.   Products of Injurious Agents

Some noxious agents, such as bacteria, may possess angiogenic activity, but most injurious agents are unlikely to stimulate vascular endothelial cell proliferation themselves. Although evidence to implicate specific bacterial products in corneal angiogenesis is not yet available, a bacterial product of *Bartonella bacilliformis* has been shown to stimulate the proliferation of human vascular endothelial cells and to increase the production of tissue plasminogen activity by these cells *in vitro*. This factor also stimulates the formation of new blood vessels in a subcutaneous wound healing assay (272).

Chapter 9

NATURE OF ANGIOGENIC FACTOR

It is generally accepted that directional growth occurs from a region where a chemical substance is in low concentration toward a zone containing this substance in high concentration. In 1893, Loeb (477) first postulated that some trophic chemical might be responsible for vascularization in general. In 1907 Goldman (298) provided the first clear suggestion that neoplasms produced an angiogenic agent, when he noted that neovascularization could easily be traced to regions in which the tumor had not yet advanced and he stated that "the impetus which gives rise to the proliferation of blood vessels emanates from the invading cell". That an angiogenic substance might be liberated by nonperfused or ischemic retina was proposed by Michaelson in 1948 (534). Using hamster cheek pouches Greenblatt and Shubik (320) provided convincing evidence that tumors produce a diffusible angiogenic factor in 1968.

Numerous investigators have provided evidence for, or isolated putative promoters of angiogenesis from normal plasma on serum (685), as well as normal tissues [brain (109,111,166,207,382,405,474,475,487,488,490,753), cartilage (413), choroid (481), corpus luteum (309,318,620), epidermis (814), fetal tissues (220), kidney (236,238,747), ovary (714), pineal gland (747), pituitary gland (207,270,305), placenta (553,628), platelets (141,167), retina (37,40,45,131,164-166,215,288,324,409), retinal pigment epithelium (481), salivary glands (34,357), vitreous (131,132)] as well as from various neoplasms (7,222,223,241,245,405,411,436, 438,599,668,683,686,687,795), myocardial infarcts (437), synovial fluid from individuals with various types of arthritis (osteoarthritis, rheumatoid arthritis, ankylosing spondylitis) (100), wound fluid (42), diabetic eyes (481), components of inflammatory and immunological reactions (185-187,702), such as macrophages (39,607), polymorphonuclear leukocytes (655), stimulated lymphocytes (570), fibrin degradation products (758,759). 3T3-adipocytes have also been shown to secrete factors which enhance angiogenesis in vivo and endothelial cell chemotaxis and proliferation in vitro (123,124,516) and stimulate protease release by vascular endothelial cells in culture (323). These biologically active substances have influenced the migration (including heparin, copper ions, prostaglandins), and/or mitotic activity of vascular endothelial cells, inflammatory cells, as well as other nonvascular cells.

Several growth factors have been implicated in angiogenesis and the neovascularization provoked by intracorneal inoculations has been used to support the putative angiogenic activity of fibroblast growth factor (55,306,314,686,754), epidermal growth factor (55,306,792), transforming growth factor-β (224), and vascular permeability factor (154).

There can be little doubt that diffusible factors initiate directional capillary growth in many normal and diseased tissues. While conclusive evidence concerning the identity of the initiator of corneal angiogenesis remains unknown several putative angiogenic factors have been isolated from other tissues: The polypeptide growth factors include the heparin binding growth factor (HBGF) family of proteins (110,241,317,472,475,476,752). Other growth factors implicated in angiogenesis include α transforming growth factor (674), transforming growth factor-β (635,715), angiogenin (78,86,173,223,549,685,727), epidermal growth factor (306,514,674) as well

45

as nicotinamide and a nicotinamide containing complex (436) and α-retina-derived growth factor (37,166).

Heparin Binding Growth Factors

The heparin binding growth factor (HBGF) family consists of structurally related polypeptides, whose genes have been cloned and sequenced: HBGF-1 (acidic fibroblast growth factor) (aFGF) (85,207,208,627,754), HBGF-2 (basic fibroblast growth factor) (bFGF) (2,207,473,486,547,631,679,752), HBGF-3 (int-2), HBGF-4 (hst/KS3), and HBGF-5 (fibroblast growth factor 5, FGF-5). HBGF-3, HBGF-4 and HBGF-5 are oncogene products. The heparin binding growth factors also include endothelial cell growth factor α and endothelial cell growth factor β (located on human chromosome 5) (109,111,382,483,749,760) and eye derived growth factor II (158,673). Both aFGF and bFGF are angiogenic in the CAM assay (207,754). The primary structure of several other endothelial cell mitogens [hepatoma-derived growth factor (HDGF) (411), chondrosarcoma-derived endothelial cell growth factor (40)] contain peptide sequences similar to those in basic fibroblast growth factor. Since fibroblast growth factors lack a precursor containing the extra peptide at the extreme amino terminal that acts as a "signal" to direct the ribosome to the rough endoplasmic reticulum (2,382) and appear to be located predominantly within cells, the question of how they are released from cells to have their putative angiogenic effect remains problematical. It is noteworthy that tumor necrosis factor (TNF) as well as interleukin 1, which has 30% amino acid sequence similarity with aFGF (754) also lack a signal peptide sequence. Perhaps they are only liberated following cell disruption as after cell injury. Nevertheless, extracellular heparin binding growth factors are tightly adsorbed to the extracellular matrix presumably because of their affinity for heparin-like glycosaminoglycans. HBGF-1 and HBGF-2 modulate the expression of endothelial cell derived proteases (322,323,489,553,562,681) that are believed to be important in the extracellular matrix degradation that accompanies angiogenesis. The observation of an abundance of HBGF in the extracellular matrix without a concomitant pronounced angiogenesis response has given rise to the concept that angiogenic activity may be dependent upon "hydrolytic extraction" from sites of attachment (243,383,411,755). Because heparin-binding growth factors are present within tissues and do not normally evoke angiogenesis, a control mechanism is essential if this angiogenic factor is to have physiological significance. Several lines of evidence are consistent with the concept that heparin-binding growth factors are released by hydrolytic enzymes (38,243,332,631,711,778,779). In the cornea polymorphonuclear leukocytes, mononuclear phagocytes and other leukocytes clearly have the capability of performing this function. Heparin protects the HBGF family from proteolytic modification (243,383,411).

HBGF-1 (acidic fibroblast growth factor) (aFGF)

Acidic fibroblast growth factor, which has a more restricted tissue distribution than basic fibroblast growth factor, has been implicated in angiogenesis. Heparin binding growth factor-1 adsorbed to a collagen matrix (755) or polytetrafluoroethylene fibers coated with collagen induces site directed new vessel formation (756). Receptors for bovine brain derived growth factor (a HPGF-1) have been found on aortic endothelial cells (368). The gene for this growth factor is localized on the short arm of human chromosome 5 (382). By *in situ* hybridization corneal epithelial cells in the rat have been found to normally express the aFGF gene (575), but thus far evidence to implicate this growth factor in corneal angiogenesis is weak, and solely based on the induction of neovascularization in the rabbit cornea following its intracorneal instillation (474) However, aFGF has domains with amino acid sequence similarity to interleukin 1 which has been implicated in corneal angiogenesis (754).

HBGF-2 (basic fibroblast growth factor) (bFGF)

The gene for bFGF is situated on human chromosome 4 (310,531). Basic fibroblast growth factor (bFGF) is a potent promoter of angiogenesis in the CAM (207) and when innoculated into the cornea (306,827). Moreover, this growth factor is a mitogen and chemoattractant for endothelial cells (and fibroblasts) (103,253,553) and stimulates the production of proteases, such as collagenase, plasminogen and plasminogen activator (103,169,553,663,821), which are capable of degrading the extracellular matrix. Receptors for endothelial cell growth factor have been found on capillary endothelial cells (261) and bFGF binds to the normal microvasculature at the corneoscleral limbus as well as to newly formed corneal blood vessels, supporting the view that vascular endothelium has receptors for bFGF (713) Further support for a role of bFGF in angiogenesis is the observation that blood vessels are reduced in the granulation tissue of subcutaneous wounds exposed to a neutralizing antibody directed against bFGF (97). Antibodies to bFGF block the ability of bovine aortic endothelial cells to migrate from the edge of a denuded area in *in vitro* endothelial wounding studies (663). Monoclonal antibodies directed against bFGF inhibit the angiogenic response to bFGF in the rat kidney capsule model of angiogenesis (623). Recombinant bFGF increases the cellularity of granulation tissue (96). The angiogenic effect of intracorneal bFGF has been shown to be enhanced by the addition of gangliosides (mono- and trisialotetraesosyl ganglioside sodium salts) (827). The mechanism by which bFGF exerts its angiogenic effect is incompletely understood and may involve protein kinase C (PKC) (83,170,552). PKC is activated by bFGF at mitogenic concentrations (83,552). The mitotic and chemotactic effects of bFGF on cultured vascular endothelium (bovine cerebral cortex capillaries) are both suppressed by staurosporine, a potent inhibitor of PKC as well as by prolonged treatment with phorbol esters that down regulate PKC (170). The question of whether bFGF contributes to corneal angiogenesis remains unanswered. In the normal rat cornea bFGF gene expression has not been detected by *in situ* hybridization (575), but bFGF is known to be sequestered normally within Descemet's membrane (242,243,781).

Epidermal-Like Growth Factors

Epidermal Growth Factor (EGF)

Epidermal growth factor (urogastrone) is a small polypeptide hormone (145) that is encoded by a gene on the large arm of chromosome 7. It elicits biologic responses by binding to a specific transmembrane surface receptor containing a cytoplasmic protein tyrosine kinase (281,762). EGF stimulates the proliferation of several cell types including corneal fibroblasts (keratocytes) (57). The EGF receptor has extensive amino acid sequence similarities with the product of the cellular oncogene, c-erb B (373). EGF is reported to promote angiogenesis in the hamster cheek pouch biossay (614), to stimulate DNA synthesis in cultured bovine pulmonary artery endothelial cells (674), to stimulate bovine retinal capillary cell migration and proliferation (514) and to elicit angiogenesis within the cornea following its implantation into that tissue (306).

Transforming Growth Factor Alpha

Transforming growth factor-α, a peptide structurally related to epidermal growth factor and secreted by a wide variety of tumor cell lines and virally

47

transformed cells, promotes angiogenesis in the hamster cheek pouch biossay and stimulates DNA synthesis in cultured bovine pulmonary artery endothelial cells (674).

Other Growth Factors

Angiogenin

Angiogenin, a 14-kDa cationic single-chain polypeptide first isolated from a human colon adenocarcinoma cell conditioned medium (223), induces an angiogenic response on the chick CAM (173,223) and the rabbit cornea (223). Angiogenin which has 35% amino acid sequence similarity with human pancreatic ribonuclease A (727), activates phospholipase C in cultured vascular endothelial cells and aortic smooth muscle cells leading to an increased cellular 1,2 diacylglycerol and inositol triphosphate (78,549).

Endothelial Cell Stimulating Angiogenesis Factor

Endothelial cell stimulating angiogenesis factor (ESAF) is a low molecular weight factor that has been isolated from tumors and a variety of other tissues (99,100,198,671,747,795). It is active in several biological systems. ESAF is often bound to carriers of a higher molecular weight (198). ESAF activates skin fibroblast procollagenase in a dose dependent manner and this property provides a technique for assaying ESAF quantitatively (796). ESAF has no apparent effect on aortic endothelial cells or fibroblasts and seems to act specifically on capillary endothelial cells (198,671).

Platelet Derived Growth Factor (PDGF)

Platelet derived growth factor, is a dimeric protein composed of 2 polypeptide chains (A and B). The genes encoding for the A and B chains are on human chromosomes 7 and 12 respectively. PDGF, which has extensive amino acid sequence similarity with the c-*sis* oncogene product, is mitogenic for vascular endothelial cells (165,826).

Platelet-Derived Endothelial Cell Growth Factor (ENDO-GF, PD-ECGF)

A growth factor isolated from platelets known as platelet-derived endothelial cell growth factor differs from PDGF. This mitogen, which does not bind to heparin, stimulates vascular endothelial cell proliferation and migration in culture (378,408). In contrast to almost all other growth factors the target cell for this growth factor seems to be specifically the vascular endothelium.

Transforming Growth Factor Beta

The family of polypeptide growth factors designated transforming growth factor beta (TGF-β) include 4 different but related compounds: TGF-β_1, TGF-β_2, TGF-β_3 and TGF-β_4 (634). TGF-β, which was originally discovered in conditioned medium from virus transformed mouse cells, is a 25 kD homodimeric peptide consisting of 112 amino acids. It is synthesized as a polypeptide precursor containing 390 amino acids. Platelets, which potentially contribute to angiogenesis, have the highest concentration of TGF-β (22,634,643). For details about these factors the reader should refer to a review on this growth factor by Roberts and colleagues (633). TGF-β can either inhibit or stimulate cell growth depending on both the cell type and the other growth factors present. TGF-β conceivably plays a role in angiogenesis of the cornea and other tissues, but if this is so the role is complex and indirect. TGF-β leads to new vessel formation following injection into the cornea (224) when injected subcutaneously in newborn mice (635), and on the CAM (634). The induced corneal angiogenesis is associated with a prominent inflammatory cell infiltrate (224). TGF-β is not chemotactic for cultured capillary endothelial cells (177), but it promotes the organization of endothelial cells into tubelike structures in three-dimensional cultures in collagen gels (492,634). However, this process is inhibited by TGF-β under certain conditions (560). However, it probably exerts this effect through some secondary mechanism as TGF-β inhibits DNA synthesis in endothelial cells, as well as endothelial cell proliferation and migration in culture (257,346,634,635). TGF-β_1 is more active (approximately 100-fold) in inhibiting the growth of endothelial cells *in vitro* than TGF-β_2 (384). Other properties attributed to TGF-β include an increase in fibronectin synthesis, and a decreased secretion of proteases by both endothelial cells and fibroblasts. TGF-β also regulates the synthesis by endothelial cells of PDGF. TGF-β is a potent chemoattract for peripheral blood monocytes (785,811), which have high affinity receptors for this growth factor consistent with their extreme sensitivity to this cytokine (785). Monocytes stimulated by TGF-β manifest augmented gene expression for several growth factors (interleukin 1, PDGF, FGF and TNF) (785,786,811). TGF-β also induces monocytes to synthesize and secrete the mature TGF-β peptide extracellularly (21) and diminishes the synthesis of both urokinase- and tissue-type plasminogen activator by cultured bovine capillary endothelial that are induced by bFGF (657). TGF-β_2 antagonizes the mitogenic effect of FGF and PDGF on vascular endothelium (362,634).

Vascular Permeability Factor

Vascular permeability factor (VPF) is a 40 kD protein originally isolated from conditioned medium of a guinea pig tumor. Without damaging the endothelium it increases vascular pemeability rapidly (within 5 minutes) and transiently (diminished by 20 minutes) (154). The action of VPF appears to be distinct from other known vasoactive agents and it does not degranulate mast cells. VPF promotes the growth of new blood vessels when administered into rat corneas and healing rabbit bone grafts (154). VPF stimulates thymidine incorporation and growth by bovine aortic and capillary endothelial cells as well as human umbilical vein endothelial cell, but not by mouse 3T3 fibroblasts or bovine smooth muscle cells. ^{125}I-VPF binds specifically and with high affinity to cultured vascular endothelium. Antibodies to the N-terminal peptide of VPF block the permeability enhancement of VPF (680). The cDNA sequence of VPF from a human histiocytic lymphoma cell line U937 has been shown to code for a 189-amino acid polypeptide that is similar in structure to the B chain of platelet derived growth factor (PDGF-B) and other PDGF-B-related proteins (403). VPF has a similar amino acid sequence to a heparin-binding growth factor specific for vascular endothelial cells and designated vascular endothelial growth factor (VEGF) (461).

Miscellaneous

An endothelial cell growth factor that is angiogenic in the chick CAM assay has been purified from conditioned medium of a pituitary cell line (AtT-20) (601). It is reported to not stimulate the growth of other cells of the vascular system, such as vascular smooth muscle cells, or that of mesoderm and neuroectoderm derived cells.

Almost all of the putative angiogenic factor have not been convincingly found to act solely and specifically on vascular endothelium. Various polypeptide growth factors (including fibroblast growth factor (306,313), macrophage derived growth factor (286), epidermal growth factor (306,514,674), platelet derived growth factor (167,826) are mitogenic for vascular endothelial cells (but also for other cells). For example, heparin binding growth factors stimulate the proliferation of capillary endothelial cells, vascular smooth muscle cells, fibroblasts, chondrocytes, glial cells, osteoblasts, corneal endothelial cells, adrenocortical and granulosa cells, melanocytes and many other cell types (207,312,329,752). To complicate the matter further some vascular endothelium, including endothelium from the microvasculature - the source of new vessels - is not responsive to FGF (171,422). EGF (117) and PDGF (805) also act on multiple cell types. Target cells for TNF and IL-1 activities include leukocytes and fibroblasts in addition to vascular endothelial cells (453). VPF is one of the very few endothelial cell growth factors claimed to not promote thymidine incorporation and the growth of non-endothelial cells (154), but the biological effect of this factor has not been thoroughly evaluated in all possible cell types. Platelet derived endothelial cell growth factor has also not been reported to effect non-endothelial cells (378,408).

Aside from the above, evidence exists to implicate products of all cell membranes. Lipids constitute a major component of plasma membranes and gangliosides within them have been implicated in angiogenesis (827). The addition of gangliosides [mono- and trisialotetraesosyl sodium salts GM1 and GT1b] to PGE_1 or bFGF, has been found to enhance the number and growth rate of newly formed capillaries following their intracorneal instillation. Removal of the ganglioside stimulation led to the disappearance of the newly formed capillaries. Removal of the ganglioside molecule nullified its stimulatory effort on angiogenesis. Neither GM1, GT1b or sialic acid were angiogenic by themselves (827). Additional evidence to implicate gangliosides in corneal angiogenesis is the finding of an elevated tissue ganglioside content prior to the neovascularization induced by PGE_1 (827). Sialic acid rich molecules, such as gangliosides, are involved in the binding of other compounds that have been implicated in angiogenesis such as fibronectin and growth factor receptors.

Chapter 10

CURRENT CONCEPT OF CORNEAL ANGIOGENESIS

Taken together, all observations suggest that corneal angiogenesis is initiated and controlled by numerous cells and molecules involved in the inflammatory response. Although the precise mechanism of corneal neovascularization remains poorly understood, many of the participants in the drama of corneal angiogenesis have been identified during the past two decades. Despite the possible existence of non-inflammatory angiogenesis in the cornea, convincing evidence to support such a variety of neovascularization in this tissue remains to be presented.

A vast body of information has now accumulated about the various cellular and humoral elements within the inflammatory response that participate in corneal angiogenesis. A considerable hiatus, however, still remains in our knowledge about the relative importance of these components in the drama of angiogenesis. Indeed, associated innocent bystanders that have nothing to do with angiogenesis have not been distinguished from active participants with certainty. Even the relative roles of the stars of the show still need to be determined. Some elements involved in the phenomenon may initiate, or be components of, relevant cascades that culminate in neovascularization.

From the many studies on corneal angiogenesis it has become apparent that the growth of blood vessels into the cornea involves a complex cascade of several concurrent and overlapping events. The complexity of the phenomenon, which includes all aspects of inflammation, makes it extremely difficult to dissect the various components in the chain of events that culminate in the formation of new vessels. The initial event that triggers pericorneal blood vessels to invade the cornea occurs early in the inflammatory response, but virtually nothing is known about its nature and little is understood about the presumed factors that stimulate vascular endothelium to invade the cornea. While one can only speculate about the fundamental events that lead to corneal vascularization, it seems likely from studies in a wide variety of biological systems that the several stimuli initiate corneal angiogenesis by different pathways and that some of them are common to angiogenic processes in other tissues.

The invasion of the cornea by blood vessels clearly depends upon the migration into the tissue of endothelial cells from pericorneal blood vessels as well as upon the ability of these cells to replicate and become aligned in an orderly manner. The message that activates vascular endothelial cells to undergo cell division, to migrate and to synthesize the various proteins that are necessary for this activity occurs soon after certain types of corneal injury. Available evidence indicates that multiple different primary messages exist and that most, if not all of, them reach the cornea during the inflammatory response.

If one extrapolates from other cellular processes one would suspect that various biologically active substances, including growth promoting factors, reach the pericorneal blood vessels from the extravascular compartment and bind to specific receptors located in the plasma membrane of the vascular endothelium, especially of post-capillary venules, and that by a variety of different signalling mechanisms the appropriate endothelial cells respond. Studies on growth factors, oncogenes, and cell replication in many systems, indicate that various growth factors are regulated by high affinity receptors on the plasma membrane of the responding cells (Fig. 14). Well characterized growth factors not only have significant regions of amino acid sequence similarity, with proteins encoded by many oncogenes, and growth factor receptors also have intrinsic protein kinase activity. While specific mechanisms of this nature remain to be elucidated for angiogenesis in general and for corneal vascularization in particular, several potential pathways are recognized in other systems (213). The biological activity of at least some endothelial cell polypeptide mitogens is mediated by a receptor present on the surface of the target cell (382) and the binding to the receptor is potentiated by heparin (109). The interaction of polypeptide growth factors with relatively high-affinity cell-surface receptors is followed by the subsequent expression of certain genes within the responsive cells. The cascade initiated by the growth factor that is responsible for mitogenic and other activity is virtually unknown, but enzymes that phosphorylate various other enzymes critical to specific cell functions (protein kinases), alterations in cyclic nucleotide metabolism, changes in inositol-3-phosphate levels and intracellular calcium levels are believed to be the mediators of the polypeptide growth factors. Following receptor interactions some growth factors are known to stimulate adenylatecyclase which leads to an increased production of cyclic adenosine 3'-5' monophosphate (cAMP) and the activation of cAMP - dependent protein kinases. Other primary messengers activate the phosphodiesterase "phospholipase C" which catalyzes the degradation of inositol-containing phospholipids.

Several lines of evidence point to the importance of protein phosphorylation in the regulation of cell proliferation by growth factors. Such phosphorylations are governed by the large family of protein kinases, which phosphorylate tyrosine or serine/threonine in the appropriate protein (331) and phosphatases which are targets for the action of growth factors. The receptors for some growth factors, including those for EGF and PDGF belong to the protein-tyrosine kinase family (331). Potentially important in angiogenesis of the cornea, as well as other tissues, are the phosphoinositol pathway and the calcium-phospholipid-dependent protein kinase C (PKC). The multifunctional PKC has been shown to play an important role in the control of cell division and signal transduction across the cell membrane. The activity of PKC is altered by several lipids, including diacylglycerol and gangliosides, which may interact directly or indirectly with the enzyme to effect the activities of enzymes phosphorylated by this kinase (213). At least in the rat acetylcholine stimulates the formation of phosphatidic acid from ^{14}C arachidonate labelled phospholipids in the cornea (613). This stimulation, which may involve muscarinic cholinergic receptors, can be completely blocked by atropine and scopolamine and partially blocked by d-tubocurarine. Muscarinic cholinergic receptors may be an important regulator of the phosphatidyl inositol (PI) cycle in corneal epithelium and thus may affect intracellular calcium mobilization (43). Tyrosine protein kinases which are located on the inner surface of the plasma membrane, phosphorylate both tyrosine and phosphatidyl inositol. The latter leads to the hydrolysis of inositol phosphates forming diacylglycerol (DG) which activates PKC. The hydrolysis of inositol phospholipids by phospholipase C generates 2 important second messengers: inositol 1,4,5-triphosphate (IP$_3$)and DG. IP$_3$ controls ionic events implicated in cell proliferation in several tissues, while DG activates a Na$^+$, H$^+$ exchange carrier increasing intracellular pH (548) leading to the activation of PKC. IP3 releases calcium from intracellular stores. Phorbol esters, which can bypass cell surface signal transduction mechanisms activate PKC in a manner analogous to DG. They appear to initiate this response in the responding cells by binding to a receptor, identified as PKC (594). Phorbol esters induce angiogenesis following intracorneal inoculation (551), in the chick CAM (551) and when added to vascular endothelial cells cultured on a collagenous matrix (543,544). However, the PKC activator β-phorbol

12,13-dibutyrate (PDBu) suppresses bovine capillary endothelial cell proliferation and DNA synthesis in response to a tumor derived heparin binding growth factor (human hepatoma-derived growth factor) (181). PDBu has no effect on the proliferation of bovine aortic endothelial cells and is mitogenic for bovine aortic smooth muscle and BALB/c 3T3 fibroblasts (181). The inhibition of human-derived growth factor-stimulated bovine capillary endothelial cell proliferation appears to be mediated through PKC (181). The mechanism by which angiogenesis is induced by some phorbol esters is incompletely understood, but since these tumor promoters increase the secretion of collagenase and plasminogen activator by capillary endothelial cells (322), the angiogenic response is probably enhanced by the increased ability of endothelial cells to degrade and hence invade the adjacent extracellular matrix. Phorbol esters that activate PKC (12-0-tetradecanoyl phorbol-13-acetate and phorbol 12,13-didecanote) stimulate angiogenesis in the cornea and on the chick CAM, but 4α-phorbol 12,13-didecanoate which is inactive as a tumor promoter, neither activates PKC, nor stimulates angiogenesis (551). The concept that PKC plays a role in angiogenesis is also supported by the evidence discussed on page 46 under bFGF.

Future studies on corneal angiogenesis will hopefully lead to a clearer understanding of the factors and of their relative contributions involved in different pathologic states.

REFERENCES

1. Abercrombie, M. Localized formation of new tissue in an adult mammal. Sym. Soc. Exp. Biol. 11:235-254, 1957.

2. Abraham, J.A., Mergia, A., Whang, J.L., Tumolo A., Friedman, J., Hjerrild, K.A., Gospodarowicz, D., and Fiddes, J.C. Nucleotide sequence of a bovine clone encoding the angiogenic protein, basic fibroblast growth factor. Science 233:545-548, 1986.

3. Adams, S.S., Burrows, C.A., Skeldon, N., and Yates, D.B. Inhibition of prostaglandin synthesis and leukocyte migration by flurbiprofen. Curr. Med. Res. Opin. 5:11-16, 1977.

4. Ahuja, O.P., and Nema, H.V. Experimental corneal vascularization and its management. Am. J. Ophthalmol. 62:707-710, 1966.

5. Albanese, A.A. Corneal vascularization in rats on a tryptophane deficient diet. Science 101:619, 1945.

6. Albanese, A.A., and Buschke, W. On cataract and certain other manifestations of tryptophane deficiency in rats. Science 95:584-586, 1942.

7. Alderman, E.M., Lobb, R.R., Fett, J.W., Riordan, J.F., Bethune, J.L., and Vallee, B.L. Angiogenic activity of human tumor plasma membrane components. Biochemistry 24:7866-7871, 1985.

8. Alessandri, G., Raju, K., and Gullino, P.M. Mobilization of capillary endothelium in vitro induced by effectors of angiogenesis in vivo. Cancer Res. 43:1790-1797, 1983.

9. Alessandri, G., Raju, K.S., and Gullino, P.M.: Interaction of gangliosides with fibronectin in the mobilization of capillary endothelium. Possible influence on the growth of metastasis. Invasion Metastasis 6:145-165, 1986.

10. Andrade, S.P., Fan, F.-P.D., and Lewis, G.P. Quantitative in-vivo studies on angiogenesis in a rat sponge model. Brit. J. Exp. Pathol. 68:755-766, 1987.

11. Arentsen, J.J. Corneal neovascularization in contact lens wearers. Int. Ophthalmol. Clin. 26:15-23, 1986.

12. Arnold, F., West, D., Schofield, P.F., and Kumar, S. Angiogenic activity in human wound fluid. Int. J. Clin. Exp. Microirc. 5:381-386, 1987.

13. Aronson, S.B., Fish, M.B., Pollycove, M., and Coon, M.A. Altered vascular permeability in ocular inflammatory disease. Arch. Opththalmol. 85:455-466, 1971.

14. Ashino-Fuse, H., Takano, Y., Oikawa, T., Shimamura, M., and Iwaguchi, T. Medroxyprogesterone acetate, an anti-cancer and anti-angiogenic steroid, inhibits the plasminogen activator in bovine endothelial cells. Int. J. Cancer 44:859-864, 1989.

15. Ashton, N.: Retinal vascularization in health and disease. Proctor Award lecture of the association for research in ophthalmology. Amer. J. Ophthalmol. 44:(Part 2) 7-17, 1957.

16. Ashton, N. Corneal vascularization. In The Transparency of the Cornea, (Ed. Duke-Elder, S., and Perkins, E.S.), Oxford, Blackwell Scientific Publications, Ltd., 1960, 131-145.

17. Ashton, N.: Neovascularization in ocular disease. Trans. Ophthalmol. Soc. UK. 81:145-161, 1961.

18. Ashton, N., and Cook, C. Mechanism of corneal vascularization. Br. J. Ophthalmol. 37:193-209, 1953.

19. Ashton, N., and Cook, C. Direct observation of the effect of oxygen on developing vessels: preliminary report. Brit. J. Ophthalmol. 38:433-440, 1954.

20. Ashton, N., Cook, C., and Langham, M. Effect of cortisone on vascularization and opacification of the cornea induced by alloxan. Br. J. Ophthalmol. 35:718-724, 1951.

21. Assoian, R.K., Fleurdelys, B.E., Stevenson, H.C., Miller, P.J., Madtes, D.K., Raines, E.W., Ross, R., and Sporn, M.B. Expression and secretion of type β transforming growth factor by activated human macrophages. Proc. Natl. Acad. Sci. USA 84:6020-6024, 1987.

22. Assoian, R.K., Komoriya, A., Meyers, C.A., Miller, D.M., and Sporn, M.B. Transforming growth factor-β in human platelets. J. Biol. Chem. 258:7155-7160, 1983.

23. Auerbach, R. Angiogenesis-inducing factors: a review. In Lymphokines: A forum for Immunoregulatory Cell Products, Volume 4, Pick, E. and Landy, M. editors, New York, Academic Press, pp. 69-88, 1981.

24. Auerbach, R. Discussion of Fraser R.A., Simpson, J.G. Role of mast cells in experimental tumour angiogenesis. In Development of the Vascular System, Ciba Foundation Symposium 100, edited by Nugent J, O'Connor M, p 120-131. London, Pitman Books, 1983.

25. Auerbach, R., Arensman, R., Kubai, L., and Folkman, J. Tumor-induced angiogenesis: Lack of inhibition by irradiation. Int. J. Cancer 15:241-245, 1975.

26. Auerbach, R., Kubai, L., and Sidky, Y. Angiogenesis induction by tumors, embryonic tissues and lymphocytes. Cancer Res. 36:3435-3440, 1976.

27. Auerbach, R., and Sidky, Y.A.: Nature of the stimulus leading to lymphocyte-induced angiogenesis. J. Immunol. 123:751-754, 1979.

28. Augstein. Gefäss-Studien an der Hornhaut und Iris: A. Hornhaut-Gefässe. Ztschr. f. Augenh. 8:317, 1902.

29. Ausprunk, D.H., Boudreau, C.L., and Nelson, D.A: Proteoglycans in the microvasculature. I. Histochemical localization in microvessels of the rabbit eye. Amer. J. Path. 103: 353-366, 1981.

30. Ausprunk, D.H., Boudreau, C.L., and Nelson, D.A. Proteoglycans in the microvasculature. II. Histochemical localization in proliferating capillaries of the rabbit cornea. Amer. J. Path. 103: 367-375, 1981.

31. Ausprunk, D.H., Falterman, K., and Folkman, J. The sequence of events in the regression of corneal capillaries. Lab. Invest. 38:284-294, 1978.

32. Ausprunk, D.H., and Folkman, J. Migration and proliferation of endothelial cells in preformed and newly formed blood vessels during tumor angiogenesis. Microvasc. Res. 14:53-65, 1977.

33. Ausprunk, D.H., Knighton, D.R., and Folkman, J. Vascularization of normal and neoplastic tissues grafted to the chick chorioallantois. Am. J. Pathol. 79:597-618, 1975.

34. Azizkhan, R.G., Azizkhan, J.C., Zetter, B.R., and Folkman, J. Mast cell heparin stimulates migration of capillary endothelial cells in vitro. J. Exp. Med. 152:931-944, 1980.

35. Bachsich, P., and Riddell, W.J.B. Structure and nutrition of the cornea, cartilage and Wharton's Jelly. Nature 155:271, 1945.

36. Bachsich, P. and Wyburn, G.M. The significance of the mucoprotein content on the survival of homografts of cartilage and cornea. Proc. Roy. Soc. Edin. 62:321-327, 1947.

37. Baird, A., Esch, F., Gospodarowicz, D., and Guillemin, R. Retina- and eye-derived endothelial cell growth factors. Partial molecular characterization and identity with acidic and basic fibroblast growth factors. Biochemistry 24:7860-7865, 1985.

38. Baird, A., and Ling, N. Fibroblast growth factors are present in the extracellular matrix produced by endothelial cells in vitro: Implications for a

role of heparinase-like enzymes in the neovascular response. <u>Biochem. Biophys. Res. Comm.</u> 142:428-435, 1987.

39. Baird, A., Mormède, P., and Böhlen, P. Immunoreactive fibroblast growth factor in cells of peritoneal exudate suggests its identity with macrophage-derived growth factor. <u>Biochem. Biophys. Res. Commun.</u> 126:358-364, 1985.

40. Baird, A., Mormède, P., and Böhlen, P. Fibroblast growth factor (FGF) in chondrosarcoma: inhibition of tumor growth with anti-FGF antibodies. <u>J. Cell. Biochem.</u> Suppl. 9A, p. 141, 1985.

41. Balazs, E.A., and Darzykiewicz, Z. The effect of hyaluronic acid on fibroblasts, mononuclear phagocytes and lymphocytes. In <u>Biology of the Fibroblast</u> (E. Kulonen and J. Pikkarainen, Eds.), Academic Press, New York, 1972, pp. 237-252.

42. Banda, M.J., Knighton, D.R., Hunt, T.K., and Werb, Z. Isolation of a nonmitogenic angiogenesis factor from wound fluid. <u>Proc. Natl. Acad. Sci. USA</u> 79:7773-7777, 1982.

43. Baratz, K.H., Proia, A.D., Klintworth, G.K., and Lapetina, E.G. Cholinergic stimulation of phosphatidylinositol hydrolysis by rat corneal epithelium in vitro. <u>Curr. Eye Res.</u> 6:691-701, 1987.

44. Barnhill, R.L., and Ryan, J. Biochemical modulation of angiogenesis in the chorioallantoic membrane of the chick embryo. <u>J. Invest. Dermatol.</u> 81:485-488, 1983.

45. Barritault, D., Plouët, J., Courty, J., and Courtois, Y. Purification, characterization and biological properties of the eye-derived growth factor from retina. Analogies with brain-derived growth factor. <u>Neurosci. Res.</u> 8:477-490, 1982.

46. Baserga, R. <u>Multiplication and Division in Mammalian Cells</u>. New York: Marcel Dekker, 1976.

47. Baserga, R., and Malamud, D. <u>Modern Methods in Experimental Pathology. Autoradiography. Techniques and Application.</u> Harper and Row, New York, 1969.

48. Baum, J.L., and Martola, E.L. Corneal edema and corneal vascularization. <u>Am. J. Ophthalmol.</u> 65:881-884, 1968.

49. Bazan, H.E.P., Birkle, D.L., Beuerman, R.W., and Bazan, N.G.: Inflammation induced stimulation of the synthesis of prostaglandins and lipoxygenase reaction products in rabbit cornea. <u>Curr. Eye Res.</u> 4:175-179, 1985.

50. Bazan, H.E.P., Birkle, D.L., Beuerman, R., and Bazan, N.G. Cryogenic lesion alters the metabolism of arachidonic acid in rabbit cornea layers. <u>Invest. Ophthalmol. Vis. Sci.</u> 26:474-480, 1985.

51. Beck, D.W., Olson, J.J., and Linhardt, R.J. Effect of heparin, heparin fragments, and corticosteroids on cerebral endothelial cell growth <u>in vitro</u> and <u>in vivo</u>. <u>J. Neuropath. Exp. Neurol.</u> 45:503-512, 1986.

52. Bell, M.A., and Scarrow, W.G. Staining for microvascular alkaline phosphatase in thick celloidin sections of nervous tissue. Morphometric and pathological applications. <u>Microvasc. Res.</u> 27: 189-203, 1984.

53. Belsky, E., and Toole, B.P. Hyaluronate and hyaluronidase in the developing chick embryo kidney. <u>Cell Differ.</u> 12: 61-66, 1983.

54. Bender, B.L., and Jaffe, R., Carlin, B., and Chung, A.E. Immunolocalization of entactin, a sulfated basement membrane component, in rodent tissues, and comparison with GP-2 (laminin). <u>Am. J. Pathol.</u> 103: 419-426, 1981.

55. BenEzra, D. Neovasculogenic ability of prostaglandins, growth factors, and synthetic chemoattractants. <u>Am. J. Ophthalmol.</u> 86:455-461, 1978.

56. BenEzra, D. Possible mediation of vasculogenesis by products of immune reaction. In A.M. Silverstein and G.R. O'Connor (Eds.), <u>Immunology and Immunopathology of the Eye</u>. New York: Masson, 1979. Pp. 315-318.

57. BenEzra, D., and Gery, I. Stimulation of keratocyte metabolism by products of lymphoid cells. Invest. Ophthalmol. Vis. Sci. 18:317-320, 1979.

58. BenEzra, D., Hemo, I., and Maftzir, G. The rabbit cornea - a model for the study of angiogenic factors. In Ocular Circulation and Neovascularization. BenEzra, D., Ryan, S.J., Glaser, D.M., and Murphy, R.P. (Eds), Martinus Nijhoff/Dr. W. Junk, Dordrecht, 1987.

59. BenEzra, D., Hemo, I., and Maftzir, G. In vivo angiogenic activity of interleukins. Arch. Ophthalmol. 108:573-576, 1990.

60. Berg, J.L., Pund, E.R., Sydenstricker, V.P., Hall, W.K., Bowles, L.L., and Hock, C.W. The formation of capillaries and other tissue changes in the cornea of the methionine-deficient rat. J. Nutr. 33:271-285, 1947.

61. Berggren, L., and Lempberg, R. Neovascularization in the rabbit cornea after intracorneal injections of cartilage extracts. Exp. Eye Res. 17: 261-273, 1973.

62. Berman, M., Leary, R., and Gage, J. Evidence for a role of the plasminogen activator-plasmin system in corneal ulceration. Invest. Ophthalmol. Vis. Sci. 19:1204-1221, 1980.

63. Berman, M., Winthrop, S., Ausprunk, D., Rose, J., Langer, R., and Gage, J. Plasminogen activator (urokinase) causes vascularization of the cornea. Invest. Ophthalmol. Vis. Sci. 22:191-199, 1982.

64. Besedovsky, H., del Rey, A., Sorkin, E., and Dinarello, C.A. Immunoregulatory feedback between interleukin-1 and glucocorticoid hormones. Science 233:652-654, 1986.

65. Bessey, O.A., and Wolbach, S.B. Vascularization of the cornea of the rat in riboflavin deficiency, with a note on corneal vascularization in vitamin A deficiency. J. Exp. Med. 69:1-12, 1939.

66. Beutler, B., and Cerami, A. Cachectin and tumour necrosis factor as two sides of the same biological coin. Nature 320:584, 1986.

67. Beutler, B., Krochin, N., Milsark, I.W., Luedke, C., and Cerami, A. Control of cachectin (tumor necrosis factor) synthesis: mechanisms of endotoxin resistance. Science 232:977-980, 1986.

68. Bevilacqua, M.P. Mononuclear phagocytes and plasma fibronectin. Doctoral Thesis. SUNY, Downstate Medical Center, New York, 1982, pp. 1-172.

69. Bevilacqua, M.P., Amrani, D., Mosesson, M.W., and Bianco, C. Receptors for cold-insoluble globulin (plasma fibronectin) on human monocytes. J. Exp. Med. 153:42-60, 1981.

70. Bevilacqua, M.P., Pober, J.S., Majeau, G.R., Cotran, R.S., and Gimbrone, M.A., Jr. Interleukin I (1L-1) induces biosynthesis and cell surface expression of procoagulant activity in human vascular endothelial cells. J. Exp. Med. 160:618-623, 1984.

71. Bevilacqua, M.P., Pober, J.S., Majeau, G.R., Fiers, W., Cotran, R.S., and Gimbrone M.A., Jr. Recombinant tumor necrosis factor induces procoagulant activity in cultured human vascular endothelium: characterization and comparison with the actions of interleukin 1. Proc. Natl. Acad. Sci. USA 83:4533-4537, 1986.

72. Bevilacqua, M.P., Pober, J.S., Wheeler, M.E., Cotran, R.S., and Gimbrone, M.A., Jr. Interleukin-I acts on cultured human vascular endothelium to increase the adhesion of polymorphonuclear leukocytes, monocytes and related leukocyte cell lines. J. Clin. Invest. 76:2003-2011, 1985.

73. Bevilacqua, M.P., Pober, J.S., Wheeler, M.E., Cotran, R.S., and Gimbrone, M.A., Jr. Interleukin-I activation of vascular endothelium: effects on procoagulant activity and leukocyte adhesion. Am. J. Path. 121:394-403, 1985.

74. Bevilacqua, M.P., Stengelin, S., Gimbrone, M.A., Jr., and Seed, B. Endothelial leukocyte adhesion molecule 1: an inducible receptor for neutrophils related to complement regulatory proteins and lectins. Science 243:1160-1165, 1989.

75. Bhattacherjee, P., and Eakins, K.E. Inhibition of the prostaglandin synthetase systems in ocular tissues by indomethacin. Br. J. Pharmacol. 50:227-230, 1974.

76. Bhattacherjee, P., and Eakins, K.E. Lipoxygenase products: Mediation of inflammatory responses and inhibition of their formation. In Leukotrienes Chemistry and Biology (Eds. Charkin, L.W., Belley, D.M.). London: Academic Press, pp. 195-214, 1984.

77. Bhattacherjee, P., Kalkarni, P.S., and Eakins, K.E. Metabolism of arachidonic acid in rabbit ocular tissues. Invest. Ophthalmol. Vis. Sci. 18:172-178, 1979.

78. Bicknell, R., and Vallee, B.L. Angiogenin activates endothelial cell phospholipase C. Proc. Natl. Acad. Sci. USA 85:5961-5965, 1988.

79. Billingham, R.E., and Boswell, T. Studies on the problem of corneal homografts. Proc. R. Soc. Lond. [Biol] 141:392-406, 1953.

80. Binder, P.S. Complications associated with extended wear soft contact lenses. Ophthalmology 86:1093-1101, 1979.

81. Bito, L.Z. Absorptive transport of prostaglandins from intraocular fluids to blood: a review of recent findings. Exp. Eye Res. 16: 299-306, 1973.

82. Bito, L.Z., and Stjernschantz, J. (Eds.). The ocular effects of prostaglandins and other eicosanoids. Alan R. Liss, Inc., New York, 1989.

83. Blackshear, P.J., Witters, L.A., Girard, P.R., Kuo, J.F., and Quamo, S.M. Growth factor-stimulated protein phosphorylation in 3T3-L1 cells. J. Biol. Chem. 260:13304-13315, 1985.

84. Boggs, D.R., Athens, J.W., Cartwright, G.E., and Wintrobe, M.M. The effects of adrenal glucocorticosteroids upon the cellular composition of inflammatory exudates. Am. J. Path. 44:763-773, 1964.

85. Böhlen, P., Esch, F., Baird, A., and Gospodarowicz, D. Acidic fibroblast growth factor (FGF) from bovine brain: amino-terminal sequence and comparison with basic FGF. EMBO J. 4:1951-1956, 1985.

86. Bond, M.D., and Strydom, D.J. Amino acid sequence of bovine angiogenin. Biochemistry 28:6110-6113, 1989.

87. Bonnet, M., and Grange, J.D. Apport du laser à l'argon au traitement des vascularisations cornéennes. Bull. Soc. Ophtalmol. Fr. 1:31-32, 1977.

88. Borgeat, P., and Samuelsson, B. Arachidonic acid metabolism in polymorphonuclear leukocytes: effect of ionophore A23187. Proc. Natl. Acad. Sci. USA 76:2148-2152, 1979.

89. Bowersox, J.C., and Sorgente, N. Chemotaxis of aortic endothelial cells in response to fibronectin. Cancer Res. 42:2547-2551, 1982.

90. Bowles, L.L., Allen, L., Sydenstricker, V.P., Hock, C.W., and Hall, W.K. The development and demonstration of corneal vascularization in rats deficient in vitamin A and in riboflavin. J. Nutr. 32:19-35, 1946.

91. Bowman, C.M., Berger, E.M., Butler, E.N., Toth, K.M., and Repine, J.E. Herpes may stimulate endothelial cells to make growth-retarding oxygen metabolites. In Vitro Cell Devel. Biol. 21: 140-142, 1985.

92. Breebaart, A.C., and James-Whitte, J. Studies on experimental corneal allergy. Am. J. Ophthalmol. 48:37-47, 1959.

93. Brem, H., and Folkman, J. Inhibition of tumor angiogenesis mediated by cartilage. J. Exp. Med. 141:427-439, 1975.

94. Brem, S. The role of vascular proliferation in the growth of brain tumors. Clin. Neurosurg. 23:440-453, 1975.

95.   Brem, S., Brem, H., Folkman, J., Finkelstein, D., and Patz, A.   Prolonged tumor dormancy by prevention of vascularization in the vitreous.   Cancer Res. 36:2807-2812, 1976.

96.   Broadley, K.N., Aquino, A.M., Hicks, B., Ditesheim, J.A., McGee, G.S., Demetriou, A.A., Woodward, S.C., and Davidson, J.M. The diabetic rat as an impaired wound healing: stimulatory effects of transforming growth factor beta and basic fibroblast factor. Biotechn. Ther. 1:55-68, 1989.

97.   Broadley, K.N., Aquino, A.M., Woodward, S.C., Buckley-Sturrock, A., Sato, Y., Rifkin, D.B., and Davidson, J.M. Monospecific antibodies implicate basic fibroblast growth factor in normal wound repair. Lab. Invest. 6:571-575, 1989.

98.   Brouty-Boyé, D., and Zetter, B.R.   Inhibition of cell motility by interferon. Science 208:516, 1980.

99.   Brown, R.A., Tomlinson I.W., Hill, C.R., Weiss, J.R., Phillips, P., and Kumar, S. Relationship of angiogenesis factor in synovial fluid to various joint diseases. Annal. Rheum. Dis. 42: 301-307, 1983.

100.   Brown, R.A., Weiss, J.B., Tomlinson, I.W., Phillips, P., and Kumar, S. Angiogenic factor from synovial fluid resembling that from tumours. Lancet 1: 682-685, 1980.

101.   Brown, S.I., Wassermann, H.E., and Dunn, M.W.   Alkali burns of the cornea. Arch. Ophthalmol. 82:91-94, 1969.

102.   Brückner, A.   Klinische Studien über Hornhaut-Gefässe.   Arch. f. Augenh. 62:17-41, 1909.

103.   Buckley-Sturrock, A., Woodward, S.C., Senior, R.M., Griffin, G.L., Klagsburn, M., and Davidson, J.M.   Differential stimulation of collagenase and chemotactic activity in fibroblasts derived from rat wound repair tissue and human skin by growth factors. J. Cell. Physiol. 138:70-78, 1989.

104.   Budden, F.H. Ocular lesions of onchocerciasis. Brit. J. Ophthalmol. 46:1-11, 1962.

105.   Bugos, H.   Angiogenic and growth factors in human amino-chorion and placenta. European J. Clin. Investig. 13: 289-296, 1983.

106.   Burger, P.C., Chandler, D.B., and Klintworth, G.K.   Corneal neovascularization as studied by scanning electron microscopy of vascular casts. Lab. Invest. 48:169-180, 1983.

107.   Burger, P.C., Chandler, D.B., and Klintworth, G.K.   Experimental corneal neovascularization: biomicroscopic, angiographic, and morphologic correlation. Cornea 4:35-41, 1985/1986.

108.   Burger, P.C., and Klintworth, G.K.   Autoradiographic study of corneal neovascularization induced by chemical cautery. Lab. Invest. 45:328-335, 1981.

109.   Burgess, M., Howk, R., Burgess, W., Ricca, G.A., Ing-Ming, C., Ravera, M.W., O'Brien, S.J., Modi, W.S., Maciag, T., and Drohan, W.N. Human endothelial cell growth factor: cloning, nucleotide sequence, and chromosome localization. Science 233:541-545, 1988.

110.   Burgess, W.H., and Maciag, T. The heparin-binding (fibroblast) growth factor family of proteins. Annu. Rev. Biochem. 58:575-606, 1989.

111.   Burgess, W.H., Mehlman, T., Friesel, R., Johnson, W.V., and Maciag, T. Multiple forms of endothelial cell growth factors. J. Biol. Chem. 260:11389-11392, 1985.

112.   Burns, R.P., Beard, M.E., Weimar, V.L., and Squires, E.L.   Modification of l-tyrosine-induced keratopathy by adrenal corticosteroids. Invest. Ophthalmol. 13:39-45, 1974.

113. Buschke, W. Classification of experimental cataracts in the rat: Recent observations on cataract associated with tryptophan deficiency and with some other experimental conditions. Arch. Ophthalmol. 30:735-762, 1943.

114. Cameron, G.R. Pathology of the Cell. Springfield, Ill. Charles C. Thomas, 1951.

115. Campbell, F.W., and Ferguson, I.D. The role of ascorbic acid in corneal neovascularization Brit. J. Ophthalmol. 33:329-334, 1950.

116. Campbell, F.W., and Michaelson, I.C. Blood vessel formation in the cornea. Br. J. Ophthalmol. 33:248-255, 1949.

117. Carpenter, G., and Cohen, S. Epidermal growth factors. Annu. Rev. Biochem. 48:193-216, 1979.

118. Carrel, A. Growth-promoting function of leucocytes. J. Exp. Med. 36:385-391, 1922.

119. Carrel, A., and Ebeling, H.-H. Tréphones embryonnaires. Comptes Rendus de la Société de Biologie 89:1142-1144, 1923.

120. Carter-Dawson, L., Tanka, M., Kuwabara, T., and Bieri, J.E. Early corneal changes in vitamin A deficient rats. Exp Eye Res 30: 261-268, 1980.

121. Castellot, J.J. Jr., Addonizio, M.L., Rosenberg, R.D., and Karnovsky, M.J. Cultured endothelial cells produce a heparin-like inhibitor of smooth muscle cell growth. J. Cell Biol. 90:372-379, 1981.

122. Castellot, J.J. Jr., Kambe, A.M., Dobson, D.E., and Spiegelman, B.M. Heparin potentiation of 3T3-adipocyte stimulated angiogenesis: mechanisms of action on endothelial cells. J. Cell. Physiol. 127:323-329, 1986.

123. Castellot, J.J. Jr., Karnovsky, M.J., and Spiegelman, B.M. Potent stimulation of vascular endothelial cell growth by differentiated 3T3 adipocytes. Proc. Natl. Acad. Sci. USA 77:6007-6011, 1980.

124. Castellot, J.J. Jr., Karnovsky, M.J., and Spiegelman, B.M. Differentiation-dependendent stimulation of neovascularization and endothelial cell chemotaxis by 3T3 adipocytes. Proc. Natl. Acad. Sci. 79: 5597-5601, 1982.

125. Cavallo, T., Sade, R., Folkman, J., and Cotran, R.S. Tumor angiogenesis. Rapid induction of endothelial mitoses demonstrated by autoradiography. J. Cell Biol. 54:408-420, 1972.

126. Cavallo, T., Sade, R., Folkman, J., and Cotran, R.S. Ultrastructural autoradiographic studies of the early vasoproliferative response in tumor angiogenesis. Am. J. Pathol. 70:345-362, 1973.

127. Chan, K.Y. Release of plasminogen activator by cultured corneal epithelial cells during differentiation and wound closure. Exp. Eye Res. 42: 417-431, 1986.

128. Chang, C.-T., Chen, Y.-L., Lee, S.-H., Lue, C.-M., and Lin, M.T. The inhibition of prostaglandin E1-induced corneal neovascularization by steroid eye drops. J. Formosan Med. Assoc. 88:707-711, 1989.

129. Chantry, D., Turner, M., and Feldman, M. Regulation of interleukin 1 and tumour necrosis factor mRNA and protein by transforming growth factor beta. Lymphokine Res. 7:283, 1988.

130. Chayet, J. Contact lens-related deep stromal neovascularization. Am. J. Ophthalmol. 107:572-573, 1989.

131. Chen, C.H., and Chen, S.C. Angiogenic activity of vitreous and retinal extract. Invest. Ophthalmol. Vis. Sci. 19:596-602,1980.

132. Chen, S., and Chen, C-H. Vascular endothelial cell effectors in fetal calf retina, vitreous and serum. Invest. Ophthalmol. Vis. Sci. 23:340-350, 1982.

133. Cherry, P.M.H,, Faulkner, J.D., Shaver, R.P., Wise, J.B., and Witter, S.L. Argon laser treatment of corneal neovascularization. Ann. Ophthalmol. 5: 911-920, 1973.

134. Cherry, P.M.H., and Garner, A. Corneal neovascularization treated with argon laser. Brit. J. Ophthalmol. 60:464-472, 1976.

135. Chiarugi, V., Ruggiero, M., Porciatti, F., Vannucchi, S., and Ziche, M. Cooperation of heparin with other angiogenetic effectors. Int. J. Tiss. React. VIII:129-133, 1986.

136. Chung, S.M., Proia, A.D., Klintworth, G.K., Watson, S.P., and Lapetina, E.G. Deoxycholate induces the preferential hydrolysis polyphosphoinositides by human platelet and rat corneal phospholipase C. Biochem. Biophys. Res. Comm. 129:411-416, 1985.

137. Claman, H.N.: Corticosteroids and lymphoid cells. N. Engl. J. Med. 287:388, 1972.

138. Clark, E.R., Kirby-Smith, H.T., Rex, R.O., and Williams, R.E. Recent modifications in the method of studying living cells and tissues in transparent chambers inserted in the rabbit's ear. Anat. Rec. 47:187, 1930.

139. Clark, R.A., Stone, R.D., Leung, D.Y.K., Silver, I., Hohn, D.C., and Hunt, T.K. Role of macrophages in wound healing. Surg. Forum 27:16-18, 1976.

140. Clementi, F., and Palade, G.E. Intestinal capillaries. I. Permeability to peroxidase and ferritin. J. Cell Biol. 41: 33-58, 1969.

141. Clemmons, D.R., Isley, W.L., and Brown, M.T. Dialyzable factor in human serum of platelet origin stimulates endothelial cell replication and growth. Proc. Natl. Acad. Sci. USA 80:1641-1645, 1983.

142. Cliff, W.J. Observations on healing tissue: a combined light and electron microscopic investigation. Phil. Trans. R. Soc. (B). 246:305-325, 1963.

143. Cogan, D.G. Vascularization of the cornea: Its experimental induction by small lesions and a new theory of its pathogenesis. Arch. Ophthalmol. 41:406-416, 1949.

144. Cogan, D.G., and Kuwabara, T. Lipogenesis of cells of the cornea. Arch. Pathol. 59:453-456, 1955.

145. Cohen, S. Epidermal growth factor. In Vitro Cell. Develop. Biol. 23:239-246, 1987.

146. Collin, H.B. Corneal lymphatics in alloxan vascularized rabbit eyes. Invest. Ophthalmol. 5:1-13, 1966.

147. Collin, H.B. Ocular lymphatics. Amer. J. Optom. 43:96-106, 1966.

148. Collin, H.B. Endothelial cell lined lymphatics in the vascularized rabbit cornea. Invest. Ophthalmol. 5:337-354, 1966.

149. Collin, H.B. Lymphatic drainage of ^{131}I-albumin from the vascularized cornea. Invest. Ophthalmol. 9:146-155, 1970.

150. Collin, H.B. Ultrastructure of lymphatic vessels in the vascularized rabbit cornea. Exp. Eye Res. 10:207-213, 1970.

151. Collin, H.B. The fine structure of growing corneal lymphatic vessels. J. Pathol. 104:99-113, 1971.

152. Collin, H.B. Limbal vascular response prior to corneal vascularization. Exp. Eye Res. 16:443-455, 1973.

153. Conn, H., Berman, M., Kenyon, K., Langer, R., and Gage, J. Stromal vascularization prevents corneal ulceration. Invest. Ophthalmol. Vis. Sci. 19:362-370, 1980.

154. Connolly, D.T., Heuvelman, D.M., Nelson, R., Olander, J.V., Eppley, B.L., Delfino, J.J., Siegel, N.R., Leimgruber, R.M., and Feder, J. Tumor vascular permeability factor stimulates cell growth and angiogenesis. J. Clin. Invest. 84:1470-1478, 1989.

155. Cooper, C.A., Bergamini, M.V.W., and Leopold, I.H. Use of flurbiprofen to inhibit corneal neovascularization. Arch. Ophthalmol. 98:1102-1105, 1980.

156. Corrent, G., Roussel, T.J., Tseng, S.C., and Watson, B.D. Promotion of graft survival by photothrombotic occlusion of corneal neovascularization. Arch. Ophthalmol. 107:1501-1506, 1989.

157. Coulombre, A.J., and Coulombre, J.L. Corneal development. III. The role of the thyroid in dehydration and the development of transparency. Exp. Eye Res. 3:105-114, 1964.

158. Courty, J., Loret, C., Moenner, M., Chevallier, B., Lagente, O., Courtois, Y., and Barritault, D. Bovine retina contains three growth factor activities with different affinity to heparin: eye derived growth factor I, II, III. Biochimie 67:265-269, 1985.

159. Crabb, C.V. Endocrine influences on ulceration and regeneration in the alkali-burned cornea. Arch. Ophthalmol. 95:1866-1870, 1977.

160. Craig, S.S., DeBlois, G., and Schwartz, L.B. Mast cells in human keloid, small intestine, and lung by an immunoperoxidase technique using a murine monoclonal antibody against tryptase. Am J. Pathol. 124:427-435, 1986.

161. Crum, R., Szabo, S., and Folkman, J. A new class of steroids inhibits angiogenesis in the presence of heparin or a heparin fragment. Science 230: 1375-1378, 1985.

162. Culton, M., Chandler, D.B., Proia, A.D., Hickingbotham, D., and Klintworth, G.K. The effect of oxygen on corneal neovascularization. Invest. Ophthalmol. Vis. Sci. 31:1277-1281, 1990.

163. Cybulsky, M.I., Colditz, I.G., and Movat, H.Z. The role of interleukin-I in neutrophil leukocyte emigration induced by endotoxin. Am. J. Pathol. 124:367-372, 1986.

164. D'Amore, P.A., and Gitlin, J.D. Effect of retinal-derived growth factor on retinal capillary endothelial cells in culture. ARVO Abstracts. Invest. Ophthalmol. Vis. Sci. 24 (Suppl): 111, 1983.

165. D'Amore, P.A., Glaser, B.M., Brunson, S.K., and Fenselau, A.H. Angiogenic activity from bovine retina: partial purification and characterization. Proc. Natl. Acad. Sci. USA 78: 3068-3072, 1981.

166. D'Amore, P.A., and Klagsbrun, M. Endothelial cell mitogens derived from retina and hypothalamus: biochemical and biological similarities. J. Cell Biol. 99:1545-1549, 1984.

167. D'Amore, P.A., and Shepro, D. Stimulation of growth and calcium influx in cultured, bovine, aortic endothelial cells by platelets and vasoactive substances. J. Cell. Physiol. 92:177-184, 1977.

168. Danon, D., Goldstein, L., Marikovsky, Y., and Skutelsky, E. Use of cationized ferritin as a label of negative charges on cell surfaces. J. Ultr. Res. 38: 500-510, 1972.

169. Davidson, J., Buckley, A., Woodward, S., Nichols, W., McGee, G., and Demetriou, A. Mechanisms of accelerated wound repair using epidermal growth factor and basic fibroblast growth factor. In Growth Factors and Other Aspects of Wound Healing: Biological and Clinical Implications, edited by Barbul, A., Pineo, E., Caldwell, M., Hunt, T.K., New York, Alan R. Liss, 1988, p. 63.

170. Daviet, I., Herbert, J.M., and Maffrand, J.P. Involvement of protein kinase C in the mitogenic and chemotaxis effects of basic fibroblast growth factor on bovine cerebral cortex capillary endothelial cells. FEB 259:315-317, 1990.

171. Davison, P.M., Bensch, K., and Karasek, U.A. Isolation and growth of endothelial cells from the microvessels of the newborn human foreskin in cell culture. J. Invest. Dermatol. 75:316-321, 1980.

172. DeBault, L.E., Esmon, N.L., Smith, G.P., and Esmon, C.T. Localization of thrombomodulin antigen in rabbit endothelial cells in culture. An

immunofluorescence and immunoelectron microscope study. <u>Lab. Invest.</u> 54:179-187, 1986.

173. Denèfle, P., Kovarik, S., Guitton, J-D., Cartwright, T., and Mayaux, J-F. Chemical synthesis of a gene coding for human angiogenin, its expression in *Escherichia coli* and conversion of the product into its active form. <u>Gene</u> 56:61-70, 1987.

174. Deutsch, T.A., and Hughes, W.F. Suppressive effects of indomethacin on thermally induced neovascularization of rabbit corneas. <u>Am. J. Ophthalmol.</u> 87:536-540, 1979.

175. DiCorleto, P.E., and Bowen-Pope, D.F. Cultured endothelial cells produce a platelet-derived growth factor-like protein. <u>Proc. Natl. Acad. Sci. USA</u> 80:1919-1923, 1982.

176. Dinarello, C.A. Interleukin-I. <u>Rev. Infect. Dis.</u> 6:51-95, 1984.

177. Dinarello, C.A. An update on human interleukin-1: from molecular biology to clinical relevance. <u>J. Clin. Immunol.</u> 5:287-297, 1985.

178. Dinarello, C.A. Interleukin-1 and its biologically related cytokines. <u>Adv. Immunol.</u> 44:153-205, 1989.

179. Dixon, J.M. Ocular changes due to contact lenses. <u>Am. J. Ophthalmol.</u> 58:424-443, 1964.

180. Dixon, J.M., and Lawaczeck, E. Corneal vascularization due to contact lenses. <u>Arch. Ophthalmol.</u> 69:72-75, 1963.

181. Doctrow, S.R., and Folkman, J. Protein kinase C activators suppress stimulation of capillary endothelial cell growth by angiogenic endothelial mitogens. <u>J. Cell Biol.</u> 104:679-687, 1987.

182. Dohlman, C.H. Corneal edema and vascularization. In <u>The Cornea World Congress</u>, King, J.H., Jr., and McTigue, J.W. Butterworths, Washington pp. 80-95, 1965.

183. Duffin, R.M., Weissman, B.A., Glasser, D.B., and Pettit, T.H. Flurbiprofen in the treatment of corneal neovascularization induced by contact lenses. <u>Am. J. Ophthalmol.</u> 93:607-614, 1982.

184. Duke-Elder, S., and Leigh, A.G. Diseases of the Outer Eye, volume 8, part 2, in <u>System of Ophthalmology</u>, Duke-Elder, S. (Ed)., Kimpton, London, 1965.

185. Dvorak, H.F. Tumors: wounds that do not heal. Similarities between tumor stroma generation and wounding healing. <u>N. Engl. J. Med.</u> 315:1650-1659, 1986.

186. Dvorak, H.F., Dvorak, A.M., Manseau, E.J., Wiberg, L., and Churchill, W.H. Fibrin-gel investment associated with line 1 and line 10 solid tumor growth, angiogenesis, and fibroplasia in guinea pigs. Role of cellular immunity, myofibroblasts, microvascular damage, and infarction in line 1 tumor regression. <u>J. Natl. Cancer Inst.</u> 62:1459-1472, 1979.

187. Dvorak, H.F., Form, D.M., Manseau, E.J., and Smith, B.D. Pathogenesis of desmoplasia. I. Immunofluorescence identification and localization of some structural proteins of line 1 and line 10 guinea pig tumors and of healing wounds. <u>J. Natl. Cancer Instit.</u> 73:1195-1205, 1984.

188. Dvorak, H.F., Harvey, V.S., Estrella, P., Brown, L.F., McDonagh, J., and Dvorak, A.M. Fibrin containing gels induce angiogenesis: implications for tumor stroma generation and wound healing. <u>Lab. Invest.</u> 57:673-686, 1987.

189. Eckardt, R.E., and Johnson, L.V. Nutritional cataract and relation of galactose to appearance of senile suture line in rats. <u>Arch. Ophthalmol.</u> 21:315, 1939.

190. Efron, N. Vascular response of the cornea to contact lens wear. <u>J. Am. Optom. Assoc.</u> 58:836-846, 1987.

191. Ehlers, H. Some experimental researches on corneal vessels. <u>Acta Ophthalmol.</u> (Kbh) 5:99-112, 1927.

192.	Eisenstein, R., Goren, S.B., Shumacher, B., and Choromokos, E. The inhibition of corneal vascularization with aortic extracts in rabbits. Am. J. Ophthalmol. 88:1005-1012, 1979.

193.	Eisenstein, R., Kuettner,K.E., Neopolitan, C., Soble, L.W., and Sorgente, N. The resistance of certain tissues to invasion: III. Cartilage extracts inhibit the growth of fibroblasts and endothelial cells in culture. Am. J. Pathol. 81:337-348, 1975.

194.	Eliason, J.A. Leukocytes and experimental corneal vascularization. Invest. Ophthalmol. Vis. Sci. 17:1087-1095, 1978.

195.	Eliason, J.A. Angiogenic activity of the corneal epithelium. Exp. Eye Res. 41:721-732, 1985.

196.	Eliason, J.A., and Elliott, J.P. Proliferation of vascular endothelial cells stimulated in vitro by corneal epithelium. Invest. Ophthalmol. Vis. Sci. 28:1963-1969, 1987.

197.	Eliason, J.A., and Maurice, D.M. Angiogenesis by the corneal epithelium. Invest. Ophthalmol. Vis. Sci. 17(Suppl.):141, 1978.

198.	Elstow, S.F., Schor, A.M., and Weiss, J.B. Bovine retinal angiogenesis factor is a small molecule (molecular mass <600). Invest. Ophthalmol. Vis. Sci. 26: 74- 79, 1985.

199.	Engerman, R.L., Pfaffenbach, D., and Davis, M.D. Cell turnover of capillaries. Lab. Invest. 17: 738-743, 1967.

200.	Eppley, B.L., Doucet, M., Connolly, D.T., Heuvelman, D., and Feder, J. Enhancement of angiogenesis by bFGF in mandibular bone graft healing in the rabbit. Oral Maxillofacial Surg. 46:391-398, 1988.

201.	Epstein, R.J., Hendricks, R.L., and Harris, D.M. Photodynamic therapy for corneal neovascularization. Cornea, submitted for publication 1990.

202.	Epstein, R.J., Hendricks, R.L., and Stulting, R.D. Interleukin-2 induces corneal neovascularization in A/J mice. Cornea 1990 In Press.

203.	Epstein, R.J., and Hughes, W.F. Lymphocyte-induced corneal neovascularization: A morphologic assessment. Invest. Ophthalmol. Vis. Sci. 21:87-94, 1981.

204.	Epstein, R.J., and Stulting, R.D. Corneal neovascularization induced by stimulated lymphocytes in inbred mice. Invest. Ophthalmol. Vis. Sci. 28:1505-1513, 1987.

205.	Epstein, R.J., Stulting, R.D., Hendricks, R.L., and Harris, D.M. Corneal neovascularization: pathogenesis and inhibition. Cornea 6:250-257, 1987.

206.	Epstein, R.J., Stulting, R.D., and Rodriques, M.M. Corneal opacities and anterior segment anomalies in DBA/2 mice. Possible models for corneal elastosis and the iridocorneal endothelial (ICE) syndrome Cornea 5:95-105, 1986.

207.	Esch, F., Baird, A., Ling, N., Ueno, N., Hill, F., Denoroy, L., Klepper, R., Gospodarowiez, D., Böhlen, P.,and Guillemin, R. Primary structure of bovine pituitary basic fibroblast growth factor (FGF) and comparison with the amino-terminal sequence of bovine brain acidic FGF. Proc. Natl. Acad. Sci. USA 82:6507-6511, 1985.

208.	Esch, F., Ueon, N., Baird, A., Hill, F., Denoroy, L., Ling,N., Gospodarowicz, D., and Guillemin, R. Primary structure of bovine brain acidic fibroblast growth factor (FGF). Biochem. Biophys. Res. Commun. 133:554-562, 1985.

209.	Ey, R.C., Hughes, W.F., Bloome, M.A., and Tallman, C.B. Prevention of corneal vascularization. Am. J. Ophthalmol 66:1118-1131, 1968.

210.	Fajardo, L.F. Special report. The complexity of endothelial cells. A review. Am. J. Clin. Path. 92:241-250, 1990.

211. Fajardo, L.F., Kowalski, J., Kwan, H.H., Prionas, S.D., and Allison, A.C. Methods in Laboratory Investigation: the disc angiogenesis system. Lab. Invest. 58:718-724, 1988.

212. Falterman, K.W., Ausprunk, H., and Klein, M.D. Role of tumor angiogenesis factor in maintenance of tumor-induced vessels. Surg. Forum 27: 157-159, 1976.

213. Farooqui, A.A., Farooqui, T., Yates, A.J., and Horrocks, L.A. Regulation of protein kinase C activity by various lipids. Neurochem. Res. 13:499-511, 1988.

214. Faure, J.P. Néovaisseaux de la cornée. Revue C. bret d'Ophtalmol. 94:18-25, 1977.

215. Federman, J.L., Brown, G.C., Felberg, N.T., and Felton, S.M. Experimental ocular angiogenesis. Am. J. Ophthalmol. 89:231-237, 1980.

216. Fehrenbacher, L., Gospodarowicz, D., and Shuman, M.A. Synthesis of plasminogen activator by bovine corneal endothelial cells. Exp. Eye Res. 29:219-228, 1979.

217. Feinberg, R.N., and Beebe, D.C. Hyaluronate in vasculogenesis. Science 220: 1177-1179, 1983.

218. Felton, S.M., Brown, G.C., Felberg, N.T., and Federman, J.L. Vitreous inhibition of tumor neovascularization. Arch. Ophthalmol. 97: 1710-1713, 1979.

219. Fenselau, A. (Ed.). Oncology Overview, National Cancer Institute, Bethesda, Md., 1983.

220. Fenselau, A., and Mello, R.J. Growth stimulation of cultured endothelial cells by tumor cell homogenates. Cancer Res. 36:3269-3273, 1976.

221. Fenselau, A., Mello, R., Kletz, M., Lutty, G., Bennett, A., Thompson, D., and Patz, A. Biochemical studies on the antiangiogenic activity from bovine vitreous (Abstract). Fed. Proc. 39:1615, 1980.

222. Fenselau, A., Watt, S., and Mello, R.J. Tumor angiogenic factor. Purification from the Walker 256 rat tumor. J. Biol. Chem. 256: 9605-9611, 1981.

223. Fett, J.W., Strydom, D.J., Lobb, R.R., Alderman, E.M., Bethune, J.L., Riordan, J.F., and Vallée, B.L. Isolation and characterization of angiogenin, an angiogenic protein from human carcinoma cells. Biochemistry 24:5480-5486, 1985.

224. Fiegel, V.D., and Knighton, D.R. Transforming growth factor-beta (TGFb) causes indirect angiogenesis by recruiting monocytes. Fed. Proc. 2:A1601, 1988.

225. Fine, B.S., Fine, S., Feigen, L., and MacKeen, D. Corneal injury threshold to carbon dioxide laser irradiation. Am. J. Ophthalmol. 66:1-15, 1968.

226. Finkelstein, D., Brem, S., Patz, A., Folkman, J., Miller, S., and Ho-Chen, C. Experimental retinal neovascularization induced by intravitreal tumors. Am. J. Ophthalmol. 83:660-664, 1977.

227. Flocks, M., Tsukahara, I., and Miller, J. Mechanically induced glaucoma in animals: Preliminary perfusion and histologic studies. Am. J. Ophthalmol. 78:11-18, 1959.

228. Flower, A.J., and Blackwell, G.J. Anti-inflammatory steroids induce leiosynthesis of a phospholipase A_2 inhibitors which presents prostaglandin generation. Nature 278:456-459, 1979.

229. Folca, P.J. Corneal vascularization induced experimentally with corneal extracts. Br. J. Ophthalmol. 53: 827-832, 1969.

230. Folkman, J. Proceedings: Tumor angiogensis factor. Cancer Res. 34: 2109-9213, 1974.

231. Folkman, J. Tumor angiogenesis. Adv. Cancer Res. 19: 331-358, 1974.

232. Folkman, J. Angiogenesis: initiation and control. Ann. NY Acad. Sci. 401:212-227, 1982.

233. Folkman, J. Surgical research: a contradiction in terms? J. Surg. Res. 36:294-299, 1984.

234. Folkman, J. Tumor angiogenesis. Adv. Cancer Res. 43: 175-203, 1985.

235. Folkman, J. Regulation of angiogenesis: a new function of heparin. Biochem. Pharmacol. 34:905-909, 1985.

236. Folkman, J. How is blood vessel growth regulated in normal and neoplastic tissue? G.H.A. Clowes Memorial Award Lecture. Cancer Res. 46:467-473, 1986.

237. Folkman, J., Ausprunk, D., and Langer, R. Connective tissue: small blood vessels and capillaries. In Textbook of Rheumatology, Kelley, W.N., Harris, E.D., Jr., Ruddy, S., and Sledge, C.B. (Eds.), Volume 1, p 210, Philadelphia, Saunders, 1981.

238. Folkman, J., and Cotran, R. Relation of capillary proliferation to tumour growth. Int. Rev. Exp. Pathol. 16: 207-248, 1976.

239. Folkman, J., and Haudenschild, C. Angiogenesis in vitro. Nature (London) 288:551-556, 1980.

240. Folkman, J., and Klagsbrun, M. Tumor angiogenesis: effect on tumor growth and immunity. Fundamental Aspects of Neoplasia (Ed. Gottlieb, A.A., Plescia, O.J., and Bishop, D.H.L.). New York, Springer-Verlag, Inc., 1975.

241. Folkman, J., and Klagsbrun, M.: Angiogenic factors. Science 235:442-447, 1987.

242. Folkman, J., Klagsbrun, M., Sasse, J., Wadzinski, M., Ingber, D., and Vlodavsky, I. Storage of a heparin-binding angiogenic factor in the cornea: a new mechanism for corneal neovascularization. Invest.Ophthalmol. Vis. Sci. 28 (Suppl) 230, 1987.

243. Folkman, J., Klagsbrun, M., Sasse, J., Wadzinski, M., Ingber, D., and Vlodavsky, I. A heparin-binding angiogenic protein - basic fibroblast growth factor - is stored within basement membrane. Am. J. Pathol. 130:393-400, 1988.

244. Folkman, J., Langer, R., Linhardt, R.J., Haudenschild, C., and Taylor, S. Angiogenesis inhibition and tumor regression caused by heparin or a heparin fragment in the presence of cortisone. Science 221:719-725, 1983.

245. Folkman, J., Merler, E., Abernathy, C., and Williams, G. Isolation of a tumor factor responsible for angiogenesis. J. Exp. Med. 133: 275-288, 1971.

246. Folkman, J., Taylor, S., and Spillberg, C. The role of heparin in angiogenesis. In Development of the Vascular System, Pitman Books, London, (Ciba Foundation Symposium 100), p 132-149, 1983.

247. Folkman, J., Weisz, P.B., Joullié, M.M., Li, W.W., and Ewing, W.R. Control of angiogenesis with synthetic heparin substitutes. Science 243:1490-1493, 1989.

248. Follis, R.H., Jr., Day, H.G., and McCollum, E.V. Histological studies of the tissues of rats fed a diet extremely low in zinc. J. Nutr. 22:223-238, 1941.

249. Fons, A., García-de-Lomas, J., Noqueira, J.M., Buesa, F.J., and Gimeno, C. Histopathology of experimental Aspergillus fumigatus keratitis. Mycopathologia 101:129-131, 1988.

250. Form, D.M., and Auerbach, R. PGE_2 and angiogenesis. Proc. Soc. Exp. Biol. Med. 172:214-218, 1983.

251. Form, D.M., Pratt, B.M., and Madri, J.A.: Endothelial cell proliferation during angiogenesis. In vitro modulation by basement membrane components. Lab. Invest. 55:521-530, 1986.

252. Fournier, G.A., Lutty, G.A., Watt, S., Fenselau, A., and Patz, A. A corneal micropocket assay for angiogenesis in the rat eye. Invest. Ophthalmol. Vis. Sci. 21:351-354, 1981.

253. Fox, G.M. The role of growth factors in tissue repair. III. Fibroblast growth factor. In The Molecular and Cellular Biology of Wound Repair, edited by Clark, R.A.F., Henson, P.M., New York, Plenum, 1988, p. 265.

254. Francois, J., and Maudgal, M.C. Experimental chloroquine keratopathy. Am. J. Ophthalmol. 60:459-464, 1965.

255. Fraser, R.A., Ellis, M., and Stalker, A.L. Experimental angiogenesis in the chorioallantoic membrane. In Current Advances in Basic and Clinical Microcirculatory Research, D.H. Lewis (Ed.), Karger, Basel pp. 25-26, 1979.

256. Fraser, R.A., and Simpson, J.E. Role of mast cells in experimental tumour angiogenesis. Ciba Foundation Symposium 100. London. Pitman Books, 1983, pp. 120-131.

257. Fräter-Schröder, M., Müller, G., Birchmeier, W., and Böhlen, R. Transforming growth factor-beta inhibits endothelial cell proliferation. Biochem. Biophys. Res. Commun. 137:295-302, 1986.

258. Fräter-Schröder, M., Risau, W., Hallmann, R., Gautschi, P., and Böhlen, P. Tumor necrosis factor type α, a potent inhibitor of endothelial cell growth in vitro, is angiogenic in vivo. Proc. Natl. Acad. Sci. USA 84:5277-5281, 1987.

259. Fratkin, J.D., DeBault, J.E., and Cancilla, P.A. Platelet-induced mitogenic stimulation of mouse cerebral vascular endothelium. J. Neuropathol. Exp. Neurol. 38:313, 1979.

260. Frederick, J.L., Shimanuki, T., and DiZerega, G.S. Initiation of angiogenesis by human follicular fluid. Science 224:389-390, 1984.

261. Freisel, R., Burgess, W.H., Mehlman, T., and Maciag, T. The characterization of the receptor for endothelial cell growth by covalent ligand attachment. J. Biol. Chem. 261:7581-7584, 1986.

262. Friedenwald, J.S. Growth pressure and metaplasia of conjunctival and corneal epithelium. Doc. Ophthalmol. 56:184,1951.

263. Fromer, C.H., and Klintworth, G.K. An evaluation of the role of leukocytes in the pathogenesis of experimentally induced corneal vascularization. I. Comparison of experimental models of corneal vascularization. Am. J. Pathol. 79:537-554, 1975.

264. Fromer, C.H., and Klintworth, G.K. An evaluation of the role of leukocytes in the pathogenesis of experimentally induced corneal vascularization: II. Studies on the effect of leukocyte elimination on corneal vascularization. Am. J. Pathol. 81:531-544, 1975.

265. Fromer, C.H., and Klintworth, G.K. An evaluation of the role of leukocytes in the pathogenesis of experimentally induced corneal vascularization: III. Studies related to the vasoproliferative capability of polymorphonuclear leukocytes and lymphocytes. Am. J. Pathol. 82:157-167, 1976.

266. Frucht, J., and Zauberman, H. Topical indomethacin effect on neovascularization of the cornea and on prostaglandin E2 levels. Brit.J. Ophthalmol. 68:656-659, 1984.

267. Frühbeis, B., Zwadlo, G., Bröcker, E.-B., Schultze Osthoff, K., Hagemeier, H.-H., Topoll, H., and Sorg, C. Immunolocalization of an angiogenic factor (HAF) in normal, inflammatory and tumor tissues. Int. J. Cancer 42:207-212, 1988.

268. Furcht, L.T. Critical factors controlling angiogenesis: cell products, cell matrix, and growth factors. Lab. Invest. 55:505-509, 1986.

269. Galli, S.J. Biology of disease. New insights into "the riddle of the mast cells": microenvironmental regulation of mast cell development and phenotypic heterogeneity. Lab. Invest. 62:5-32, 1990.

270. Gambarini, A.G., and Armelin, H.A.: Purification and partial characterization of an acidic fibroblast growth factor from bovine pituitary. J. Biol. Chem. 257:9692-9697, 1982.

271. Gamble, J.R., Harlan, J.H., Klebanoff, S.J., and Vadas, M.A. Stimulation of the adherence of neutrophils to umbilical vein endothelium by human recombinant tumor necrosis factor. Proc. Natl. Acad. Sci. USA 82:8667-867, 1985.

272. Garcia, F.U., Wojta, J., Broadley, K.N., Davidson, J.M., and Hoover, R.L. Bartonella bacilliformis stimulates endothelial cells in vitro and is angiogenic in vivo. Am. J. Path. 136:1125-1135, 1990.

273. Garner A. Pathology of ocular onchocerciasis: human and experimental. Trans. R. Soc. Trop. Med. Hyg. 70:374-377, 1976.

274. Garner, A. The pathogenesis of ocular neovascularization. Bibl. Anat. 16:14-18, 1977.

275. Garner, A. Vascular Disorders. In A. Garner and G.K. Klintworth (Eds.), Pathobiology of Ocular Disease (part B). New York: Marcel Dekker, 1982. Pp. 1479-1576.

276. Garner, A.: Ocular angiogenesis Internat. Rev. Exp. Pathol. 28:249-306, 1986.

277. Gasset, A.R., and Dohlman, C.H. The tensile strength of corneal wounds. Arch. Ophthalmol. 79:595-602, 1968.

278. Geanon, J.D., Tripathi, B.J., Tripathi, R.C., and Barlow, G.H. Tissue plasminogen activator in avascular tissues of the eye: a quantitative study of the activity in the cornea, lens, and aqueous and vitreous humors of dog, calf, and monkey. Exp. Eye Res. 44:55-63, 1987.

279. Germuth, F.G., Maumenee, A.E., Senterfit, L.B., and Pollack, A.D. Immunohistologic studies on antigen-antibody reactions in the avascular cornea. I. Reactions in rabbits actively sensitized to foreign protein. J. Exp. Med. 115:919-928, 1962.

280. Gery, I., Gershon, R.K., and Waksman, B.H. Potentiation of the T-lymphocyte response to mitogens. I. The responding cell. J. Exp. Med. 136:128-142, 1972.

281. Gill, G.N., Bertics,P.J., and Santon, J.B. Epidermal growth factor and its receptor. Molecul. Cell. Endocrinol. 53:169-186, 1987.

282. Gillette, T.E., Chandler, J.W., and Greiner, J.V. Langerhans cells of the ocular surface. Ophthalmology 89:700-711, 1982.

283. Gimbrone, M.A., Jr., Cotran, R.S., Leapman, S.B., and Folkman, J. Tumor growth and neovascularization: An experimental model using the rabbit cornea. J. Natl. Cancer Inst. 52:413-427, 1974.

284. Gimbrone, M.A., Jr., and Gullino, P.M. Neovascularization induced by intraocular xenografts of normal, preneoplastic, and neoplastic mouse mammary tissues. J. Natl. Cancer Inst. 56: 305-318, 1976.

285. Gimbrone, M.A., Jr., Leapman, S.B., Cotran, R.S., and Folkman, J. Tumor angiogenesis: iris neovascularization at a distance from experimental intraocular tumors. J. Natl. Cancer Inst. 50: 219-228, 1973.

286. Gimbrone, M.A., Jr., Martin, B.M., Baldwin, W.M., Unanue, E.R., and Cotran, R.S. (1982). Stimulation of vascular cell growth by macrophage products. In "Pathobiology of the Endothelial Cell" (H.L. Nossel and H.J. Vogel, eds.), pp. 3-17, Academic Press, New York.

287. Gipson, I.K., Burns, R.P., and Wolfe-Lande, J.D. Crystals in corneal epithelial lesions of tyrosine-fed rats. Invest. Ophthalmol. 14:937-941, 1975.

288. Glaser, B.M., D'Amore, P.A., Michels, R.G., Patz, A., and Fenselau, A. Demonstration of vasoproliferative activity from mammalian retina. J. Cell Biol. 84: 298-304, 1980.

289. Glaser, B.M., Kalebic, T., Garbisa, S., Connor, T.B., and Liotta, L.A. Degradation of basement membrane components by vascular endothelial cells: role in neovascularization. In Development of the Vascular System, Ciba Foundation Symposium 100, (ed. Nugent, J., and Connor, M.) London, Pitman Books, 1983, p 150.

290. Glatt, H.J., Halperin, E.C., Smith, C.F., and Klintworth, G.K. Corneal vascularization in irradiated inbred mice reconstituted with bone marrow. Submitted for publication.

291. Glatt, H.J., and Klintworth, G.K. Quantitation of neovascularization in flat preparations of the cornea. Microvasc. Res. 31:104-109, 1986.

292. Glatt, H.J., Vu, M.T., Burger, P.C., and Klintworth, G.K. Effect of irradiation on vascularization of corneas grafted onto chorioallantoic membranes. Invest. Ophthalmol. Vis. Sci. 26:1533-1542, 1985.

293. Glenn, K.C., and Ross, R. Human monocyte-derived growth factor(s) for mesenchymal cells: activation of secretion by endotoxin and concanavalin A. Cell 25: 603-615, 1981.

294. Glowacki, J., Kaban, L.B., Murray, J.E., Folkman, J., and Mulliken, J.B. Bone implants and induced osteogenesis [letter]. Lancet, 1:452 1982.

295. Glowacki, J., Trepman, E., and Folkman, J. Cell shape and phenotypic expression in chondrocytes. Proc. Soc. Exp. Biol. Med. 172:93-98 1983.

296. Goetz, I.E., Warren, J., Estrada, C., Roberts, E., and Krause, D.N. Long-term serial cultivation of arterial and capillary endothelium from adult bovine brain. In Vitro Cell. Develop. Biol. 21: 172-180, 1985.

297. Goldfarb, R.H. Proteases in tumor invasion and metastasis. In Tumor Cell Invasion and Metastasis, pp 375-390. Eds. L.A. Liotta and L.R. Hart. Martinus Nijhoff, The Hague, 1982.

298. Goldman, E. The growth of malignant disease in man and the lower animals, with special reference to the vascular system. Proc. R. Soc. Med 1 (Surgical section): 1-13, 1907.

299. Goldsmith, H.S., Griffith, A.L., Kupferman, A., and Catsimpoolas, N Lipid angiogenic factor from omentum. JAMA 252: 2034-2036, 1984.

300. Golub, B.M., Foster, C.S., and Colvin, R.B. Distribution of fibrin, fibronectin, laminin and type IV collagen during corneal neovascularization response in guinea pigs. Rejected by Invest. Ophthalmol. Vis. Sci., 1987.

301. Gomez, D.S., Lee, A., and Friedlaender, M.H. Ia staining of Langerhans cells in the mouse cornea. ARVO Abstracts. Invest. Ophthalmol. Vis. Sci. 26(Suppl.):236, 1985.

302. Gonzalez, A.E., and Klintworth, G.K. Enhancement of corneal neovascularization in inbred mice made thrombocytopenic with rabbit-anti-mouse-platelet serum. Invest. Ophthalmol. Vis. Sci. 26(Suppl):329, 1985.

303. Goodwin, J.S., Bankhurst, A.D., and Messner, R.P. Suppression of human T-cell mitogenesis by prostaglandin. Existence of prostaglandin-producing suppression cell. J. Exp. Med. 146: 1719-1734, 1977.

304. Goren, S.B., Eisenstein, R., and Choromokos, E. The inhibition of corneal vascularization in rabbits. Am. J. Ophthalmol. 84:305-309, 1977.

305. Gospodarowicz, D. Purification of fibroblast growth factor from bovine pituitary. J. Biol. Chem. 250:2515-2520, 1975.

306. Gospodarowicz, D., Bialecki, H., and Thakral, T.K. The angiogenic activity of the fibroblast and epidermal growth factor. Exp. Eye Res. 28:501-514, 1979.

307. Gospodarowicz, D., Brown, K.D., Birdwell, C.R., and Zetter, B.R. Control of proliferation of human vascular endothelial cells. Characterization of the response of human umbilical vein endothelial cells to fibroblast growth

factor, epidermal growth factor, and thrombin. J. Cell Biol. 77:774-778, 1978.

308. Gospodarowicz, D., and Cheng, J. Heparin protects basic and acidic FGF from inactivation. J. Cell. Physiol. 128:475-484, 1986.

309. Gospodarowicz, D., Cheng, J., Lui, G.M., Baird, A., Esch, F., and Bohlen, P. Corpus luteum angiogenic factor is related to fibroblast growth factor. Endocrinology 117:2383-2391, 1985.

310. Gospodarowicz, D., Ferrara, N., Schweigerer, L., and Neufeld, G. Structural characterization and biological functions of fibroblast growth factor. Endocrine Reviews 8:95-114, 1987.

311. Gospodarowicz, D., Greenburg, G., Bialecki, H., and Zetter, B.R. Factors involved in the modulation of cell proliferation in vivo and in vitro: the role of fibroblast and epidermal growth factors in the proliferative response of mammalian cells. In Vitro 14: 85-117, 1978.

312. Gospodarowicz, D., Massoglia, S., Cheng, J., Lui, G., and Bohlen, P. Isolation of pituitary fibroblast growth factor by fast protein liquid chromatography (FPLC): partial chemical and biological characterization. J. Cell. Physiol. 122:323-332, 1985.

313. Gospodarowicz, D., Mescher, A.L., and Birdwell, C.R. Stimulation of corneal endothelial cell proliferation in vitro by fibroblast and epidermal growth factors. Exp. Eye Res. 25:75-89, 1977.

314. Gospodarowicz, D., Mescher, A.L., and Birdwell, C.R. Control of cellular proliferation by fibroblast and epidermal growth factors. Natl. Cancer Inst. Monogr. 48:109-130, 1978.

315. Gospodarowicz, D., Moran, J.S., and Braun, D.L. Control of proliferation of bovine vascular endothelial cells. J. Cell. Physiol. 91:377-385, 1977.

316. Gospodarowicz, D., Neufeld, G., and Schneigerer, L. Review: Fibroblast growth factor. Mol. Cell. Endocrinol. 46:187-204,1986.

317. Gospodarowicz, D., Neufeld, G., and Schweigerer, L. Molecular and biological characterization of fibroblast growth factor, an angiogenic factor which also controls the proliferation and differentiation of mesoderm and neuroectoderm derived cells. Cell Diff. 19:1-17, 1986.

318. Gospodarowicz, D., and Thakral, K.K. Production of a corpus luteum angiogenic factor responsible for proliferation of capillaries of neovascularization of the corpus luteum. Proc. Natl. Acad. Sci. U.S.A. 75:847-851, 1978.

319. Graymore, C., and McCormick, A. Induction of corneal vascularization with alloxan. Br. J. Ophthalmol. 52:138-140, 1968.

320. Greenblatt, M., and Shubik, P. Tumor angiogenesis: transfilter diffusion studies in the hamster by transparent chamber techniques. J. Natl. Cancer Inst. 41:111-124, 1968.

321. Grobstein, C., and Parker, G. Epithelial tubule formation by mouse metanephrogenic mesenchyme transplanted in vivo. J. Natl. Cancer Inst. 20:107-109, 1958.

322. Gross, J.L., Moscatelli, D., Jaffe, E.A., and Rifkin, D.B. Plasminogen activator and collagenase produced by cultured capillary endothelial cells. J. Cell Biol. 95:974-981, 1982.

323. Gross, J.L., Moscatelli, D., and Rifkin, D.B. Increased capillary endothelial cell protease activity in response to angiogenic stimuli in vitro. Proc. Natl. Acad. Sci. USA 80:2623-2627, 1983.

324. Gu, X.Q., Fry, G.L., Lata, G.F., Packer, A.J., Servais, E.G., Hock, J.C., and Hayreh, S.S. Ocular neovascularization: tissue culture studies. Arch. Ophthalmol. 103:111-117, 1985.

325. Gullino, P.M. Angiogenesis factor(s). In Tissue Growth Factors. Handbook of Experimental Pharmacology, Vol. 57, pp 427-449 (Ed. R. Baserga), Springer-Verlag, New York 1981.

326. Gurina, O.I. Specific endothelial bodies in growing capillaries of the rabbit cornea (In Russian). Arch. Anat. Gistolog. Embriolog. 92:28-30, 1987.

327. Haessler, F.H. A function of blood in corneal vascularization. Trans. Am. Ophthalmol. Soc. 25:412-417, 1927.

328. Hajjar, K.A., Harpel, P.C., Jaffe, E.A., and Nachman, R.L. Binding of plasminogen to cultured human endothelial cells. J. Biol. Chem. 261:11656-11662, 1986.

329. Halaban, R., Ghosh, S., and Baird, A. bFGF is the putative natural growth factor for human melanocytes. In Vitro Cell. Develop. Biol. 23:47-52, 1987.

330. Hall, W.K., Sydenstricker, V.P., Hock, C.W., and Bowles, L.L. Protein deprivation as a cause of vascularization of the cornea in the rat. J. Nutr. 32:509-524, 1946.

331. Hanks, S.K., Quinn, A.M., and Hunter, T. The protein kinase family: conserved features and deduced phylogeny of the catalytic domains. Science 241:42-52, 1988.

332. Hanneken, A., Lutty, G.A., McLeod, D.S., Robey, F., Harvey, A.K., and Hjelmeland, L.M. Localization of basic fibroblast growth factor to the developing capillaries of the bovine retina. J. Cell. Physiol. 138:115-120, 1989.

333. Harvey, P.T., and Cherry, P.M.H. Indomethacin v. dexamethasone in the suppression of corneal neovascularization. Can. J. Ophthalmol. 18:293-295, 1983.

334. Hase, S., Nakazawa, S., Tsukamoto, Y., and Segawa, K. Effects of prednisolone and human epidermal growth factor on angiogenesis in granulation tissue of gastric ulcer induced by acetic acid. Digestion 42:135-142, 1989.

335. Haye, C., Haut, J., and Sicault, R. Premiers résultats de la cryocoagulation sur la vascularisation cornéene. Bull. Soc. Ophtalmol. Fr. 67: 13-15, 1967.

336. Hayek, A., Culler, F.L., Beattie, G.M., Lopez, A.D., Cuevas, P., and Baird, A. Biochem. Biophys. Res. Commun. 147:876-880, 1987.

337. Haynes, W.L., Proia, A.D., and Klintworth, G.K. Effect of inhibitors of arachidonic acid metabolism on corneal neovascularization in the rat. Invest. Ophthalmol. Vis. Sci. 30:1588-1593, 1989.

338. Hazlett, L.D., and Berk, R.S. Heightened resistance of athymic, nude (nu/nu) mice to experimental Pseudomonas aeruginosa ocular infection. Infect. Immun. 22:926-933,1978.

339. Heine, U.I., Munoz, E.F., Flanders, K.L., Elingsworth, L.R., Lam, H.-Y., Thompson, N.L., Roberts, A.B., and Sporn, M.B. The role of transforming growth factor-β in the development of the mouse embryo. J. Cell Biol. 105:2861-2876, 1987.

340. Hemo, I., BenEzra, D., Maftzir, G., and Birkenfeld, V. Angiogenesis and Interleukins. In Ocular Circulation and Neovascularization. BenEzra, D., Ryan, S.J., Glaser, D.M., and Murphy, R.P. (Eds), Martinus Nijhoff/Dr. W. Junk, Dordrecht, 1987.

341. Hendricks, R.L., Epstein, R.J., and Tumpey, T. The effect of cellular immune tolerance to HSV-1 antigens on the immunopathology of HSV-1 keratitis. Invest. Ophthalmol. Vis. Sci. 30:105-115, 1989.

342. Henkind, P. Hyperbaric oxygen and corneal neovascularization. Lancet 2:836, 1964.

343. Henkind, P. Migration of limbal melanocytes into the corneal epithelium of guinea pigs. Exp. Eye Res. 4:42-47, 1965.

344. Herbert, J.M., Laplace, M.C., and Maffrand, J.P. Effect of heparin on the angiogenic potency of basic and acidic fibroblast growth factors in the rabbit corneal assay. Internat. J. Tiss. React. 10:133-139, 1988.

345. Herbort, C.P., Matsubara, M., Nishi, M., and Mochizuki, M. Penetrating keratoplasty in the rat: a model for the study of immunosuppressive treatment of graft rejection. Jap. J. Ophthalmol. 33:212-220, 1989.

346. Hermark, R.L., Twardzik, D.R., and Schwartz, S.M. Inhibition of endothelial regeneration by type-beta transforming growth factor from platelets. Science 233:1078-1080, 1986.

347. Herron, G.S., Werb, Z., Dwyer, K., Banda, M.J. Secretion of metalloproteinases by stimulated capillary endothelial cells. I. Production of procollagenase and prostromelysin exceeds expression of proteolytic activity. J. Biol. Chem. 261:2810-2813, 1986.

348. Hervouet, F., and Negroz, A. Derivation of conjunctival vessel and corneal ulcers. Bull. Soc. Ophtalmol. Fr. 67: 762, 1967.(In French)

349. Heuser, L.S., Taylor, S.H., and Folkman, J. Prevention of carcinomatosis and bloody malignant ascites in the rat by an inhibitor of angiogenesis. J. Surg. Res. 1984, 36(3):244-250.

350. Higgs, G.A., McCall, E., and Youlten, L.J.F. A chemotactic role for prostaglandins released from polymorphonuclear leukocytes during phagocytosis. Brit. J. Pharmacol. 53:539-546, 1975.

351. Higgs, G.A., Moncada, S., Salmon, J.A., and Seager, K. The source of thromboxane and prostaglandins in experimental inflammation. Brit. J. Pharmacol. 79:863-868, 1983.

352. Hirata, F., Schiffmann, E., Venkatasubramanian, K., Salomon, D., and Axelrod, J. A phospholipase A_2 inhibitory protein in rabbit neutrophils induced by glucocorticoids. Proc. Natl. Acad. Sci. USA 77:2533-2536, 1980.

353. Hirata, S., Matsubara, T., Saura, R., Tateishi, H., and Hirohata, K. Inhibition of in vitro vascular endothelial cell proliferation and in vivo neovascularization by low-dose methotrexate. Arthr. Rheumat. 32:1065-1073, 1989.

354. Hoban, B.P., and Collin, H.B. Effects of salicylate and steroid on neutrophil migration and corneal blood vessel growth. Am. J. Optomet. Physiol. Optics 63:271-276, 1986.

355. Hock, C.W., Hall, W.K., Pund, E.R., and Sydenstricker, V.P. Vascularization of the cornea as a result of lysine deficiency. Fed. Proc. 4:155-156, 1945.

356. Höckel, M., Sasse, J., and Wissler, J.H. Purified monocyte-derived angiogenic substance (angiotropin) stimulates migration, phenotypic changes, and "tube formation" but not proliferation of capillary endothelial cells in vitro. J. Cell. Physiol. 133:1-13, 1987.

357. Hoffman, H., McAuslan, B., Robertson, D., and Burnett, E. An endothelial growth- stimulating factor from salivary glands. Exp. Cell Res. 102: 269-275, 1976.

358. Hohfeld, K. Corneal vascularization by injection of Boviserin-Ferritin. Albrecht von Graefes Arch. Klin. Ophthalmol. 183: 226-231, 1971.

359. Honegger, H. Hornhautvascularisation: Experimentelle Unterschung über die Rolle von Hornhautquellung und Entzündung. Albrecht v. Graefes Arch. Klin. Exp. Ophthalmol. 176:239-244, 1968.

360. Hook, C.N., Brown, S.I., Iwanij, W., and Nakanishi, I. Characterization and inhibition of corneal collagenase. Invest. Ophthalmol. 10: 496-503, 1971.

361. Hoshi, H., and McKeehan, W.L. Isolation, growth requirements, cloning, prostacyclin production and life-span of human adult endothelial cells in low serum culture medium. In Vitro Cell. Devel. Biol. 22:51-56, 1986.

362. Hotta, M., and Baird, A. Endocrinology 118 (Suppl.) 289, 1986.

363. Howes, E.L., Cruse, VK, and Kwok, M.T. Mononuclear cells in the corneal response to endotoxin. Invest. Ophthalmol. Vis. Sci. 22:494-501, 1982.
364. Hoyer, I.W., de los Santos, R.P., and Hoyer, J.R. Antihemophilic factor antigen: localization in endothelial cells by immunofluorescent microscopy. J. Clin. Invest. 52:2737-2744, 1973.
365. Huang, A.J.W., Tseng, S.C.G., and Kenyon, K.R. Alteration of epithelial paracellular permeability during corneal epithelial wound healing. Invest. Ophthalmol. Vis. Sci. 31:429-435, 1990.
366. Huang, A.J.W., Watson, B.D., Hernandez, E., and Tseng, S.C. Induction of conjunctival transdifferentiation on vascularized corneas by photothrombotic occlusion of corneal neovascularization. Ophthalmology 95:228-235, 1988.
367. Huang, A.J.W., Watson, B.D., Hernandez, E., and Tseng, S.C. Photothrombosis of corneal neovascularization by intravenous rose bengal and argon laser irradiation. Arch. Ophthalmol. 106:680-685, 1988.
368. Huang, S.S., and Huang, J.S. Association of bovine brain-derived growth factor receptor with protein tyrosine kinase activity. J. Biol. Chem. 261:9568-9571, 1986.
369. Hueper, W.C., and Martin, G.J. Tyrosine poisoning in rats. Arch. Pathol. 35:685-694, 1943.
370. Hughes, W.F., Bloome, M.A., and Tallman, C.B. Prevention of corneal vascularization. Am. J. Ophthalmol. 66:1118, 1968.
371. Humes, J.L., Bonney, R.J., Pelus, J., Dahlgren, M.E., Sadowski, S.J., Kuehl, F.A., Jr., and Davies, P. Macrophages synthesize and release prostaglandins in response to inflammatory stimuli. Nature 269:149-150, 1977.
372. Hunter, J. A treatise on the blood, inflammation and gunshot wounds. Written 1792, published London, 1812.
373. Hunter, T. Oncogenes and growth control. Trends Biochem. Sci. 10:275, 1985.
374. Imre, G. The mechanism of corneal vascularization. Acta Morphol. Acad. Sci. Hung 14:99-104, 1966.
375. Ingber, D.E., Madri, J.A., and Folkman, J. A possible mechanism for inhibition of angiostatic steroids: induction of capillary basement membrane dissolution. Endocrinology 119:1768-1775, 1986.
376. Ingber, D.E., Madri, J.A., and Folkman, J. Endothelial growth factors and extracellular matrix regulate DNA synthesis through modulation of cell and nuclear expansion. In Vitro Cell. Develop. Biol. 23:387-394, 1987.
377. Inomata, H., Smelser, G.K., and Polack, F.M. Corneal vascularization in experimental uveitis and graft rejection: an electron microscopy study. Invest. Ophthalmol. 10: 840-850, 1971.
378. Ishikawa, F., Miyazono, K., Hellman, U., Drexler,H., Wernstedt, C., Hagiwara, K., Usuki, K., Takuku, F., Risau, W., and Heldrin, C-H. Identification of angiogenic activity and the cloning and expression of platelet-derived endothelial cell growth factor. Nature (London) 338:557-562, 1989.
379. Jacobson, B., Dorfman, T., Basu, P.K., and Hasany, S.M. Inhibition of vascular endothelial cell growth and trypsin activity by vitreous. Exp. Eye Res. 41:581-595, 1985.
380. Jaffe, B.M. Prostaglandins and cancer: an update. Prostaglandins 6: 453-454, 1974.
381. Jaffe, E.A. Endothelial cells and the biology of Factor VIII. N. Engl. J. Med. 296:377, 1977.
382. Jaye, M., Howk, R., Burgess, W., Ricca, G.A., Chiu, I-M., Ravera, M.W., O'Brien, S.J., Modi, W.S., Maciag, T., and Drohan, W.N. Human endothelial

cell growth factor: cloning, nucleotide sequence, and chromosome localization. Science 233:541-545, 1986.

383.   Jeanny, J.-C., Fayein, N., Moenner, M., Chevallier, B., Barritault, D., and Coutois, Y. Specific fixation of bovine brain and retinal acidic and basic fibroblast growth factors to mouse embryonic eye basement membranes. Exp. Cell Res. 171:63-75, 1987.

384.   Jennings, J.C., Mohan, S., Linkhart, T.A., Widstrom, R. Quantitation of beta 1 and beta 2 TGF in bone matrix extracts and bone cell conditioned medium using unique biological and radioreceptor assays. Endocrinology 122(Suppl):A1222, 1988.

385.   Jensen, J.A., Hunt, T.K., Scheuenstuhl, H., and Banda, M.J. Effect of lactate, pyruvate, and pH on secretion of angiogenesis and mitogenesis factors by macrophages. Lab. Invest. 54: 574-578, 1986.

386.   Johnson, L.V., and Eckardt, R.E. Rosacea keratitis and conditions with vascularization of cornea treated with riboflavin. Arch. Ophthalmol. 23:899-907, 1940.

387.   Jørgensen, K.A., and Stoffersen, E. Hydrocortisone inhibits platelet prostaglandin and endothelial prostacyclin production. Pharmacol. Res. Commun. 13:579-586, 1981.

388.   Josephson, J.E., and Caffery, B.E. Progressive corneal vascularization associated with extended wear of a silicone elastomer contact lens. Am. J. Optometr. Physiol. Optics 64:958-959, 1987.

389.   Julianelle, L.A., and Bishop, G.H. The formation and development of blood vessels in the sensitized cornea. Am. J. Anat. 58:109-125, 1936.

390.   Julianelle, L.A., and Lamb, H.D. Studies on vascularization of the cornea. V. Histological changes accompanying corneal hypersensitiveness. Am. J. Ophthalmol. 17:916-921, 1934.

391.   Julianelle, L.A., Morris, M.C., and Harrison, R.W. An experimental study of corneal vascularization. Am. J. Ophthalmol. 16:962-966, 1933.

392.   Kagonyera, G.M., George, L.W., and Munn, R. Light and electron microscopic changes in corneas of healthy and immunomodulated calves infected with Moraxella bovis. Am. J. Vet. Res. 49:386-395, 1988.

393.   Kaiser, R.J., and Klopp, D.W.: Hyperbaric air and corneal vascularization. Ann. Ophthalmol. 5: 44-47, 1973.

394.   Kaminski, M., and Auerbach, R. Angiogenesis induction by CD-4 positive lymphocytes. Proc. Soc. Exp. Biol. Med. 188:440-443, 1988.

395.   Kamiński, M.J., Bem, W., Majewski, S., Kamińska, G., and Szmurlo, A. Angiogenesis induction by xenogeneic lymphoid and non-lymphoid cells in mice. Arch. Immun. Therap. Exper. 29:521-523, 1981.

396.   Kamiński, M., Hayari, Y., Kamińska, G., Muthukkaruppan, V., Lubai, L., and Auerbach, R. Macrophage-induced neovascularization in the mouse eye in correlation with other in vivo and in vitro tests of angiogenesis. In Ocular Circulation and Neovascularization. BenEzra, D., Ryan, S.J., Glaser, D.M., and Murphy, R.P. (Eds), Martinus Nijhoff/Dr. W. Junk, Dordrecht, 1987.

397.   Kamiński, M., and Kamińska, G. Inhibition of lymphocyte-induced angiogenesis by enzymatically isolated rabbit cornea cells. Arch. Immunol. Ther. Exp. (Warsz.) 26:1079-1082, 1978.

398.   Kan, M., Kato, M., and Yamane, I. Long-term serial cultivation and growth requirements for human umbilical vein endothelial cells. In Vitro Cell.Develop. Biol. 21: 181-187, 1985.

399.   Karasek, E., and Heder, G. The rabbit cornea model for detecting neovascularization effects. Z. Versuchstierkd. 23:59-66, 1981.

400.   Kass, M.A., and Holmberg, N.J. Prostaglandins and thromboxane synthesis by microsomes of the rabbit ocular tissue. Invest. Ophthalmol. Vis. Sci. 18:166-171, 1979.

401.   Katz, H.R., Aizuss,D.H., and Mondino, B.J. Inhibition of contact lens-induced corneal neovascularization in radial keratotomized rabbit eyes. Cornea 3:65-72, 1984.

402.   Katz, H.R., Duffin, R.M., Glasser, D.B., and Pettit, T.H. Complications of contact lens wear after radial keratotomy in an animal model. Amer. J. Ophthalmol. 94:377-382, 1982.

403.   Keck, P.J., Hauser, S.D., Krivi, G., Sanzo, K., Warren, T., Feder, J., and Connolly, D.T. Vascular permeability factor, an endothelial cell mitogen related to PDGF. Science 246:1309-1312, 1989.

404.   Keegan, A., Hill, C., Kumar, S., Phillips, P., Schor, A., and Weiss, J. Purified tumour angiogenesis factor enhances proliferation of capillary but not aortic endothelial cells in vitro. J. Cell Sci. 55: 261-276, 1982.

405.   Kellet, J.G., Tanaka, T., Rowe, J.M., Shiu, R.P.C., and Friesen, H.G. The characterization of growth factor activity in the brain. J. Biol. Chem. 256:54-58, 1981.

406.   Kessler, D.A., Langer, R.S., Pless, N.A., and Folkman, J. Mast cells and tumor angiogenesis. Int. J. Cancer 18:703-709, 1976.

407.   Khodadoust, A.A. The allograft rejection reaction: the leading cause of late failure of clinical corneal grafts. In Ciba Foundation Symposium 15: Corneal Graft Failure, Porter, R., and Knight, J. (Ed.), Elsevier/North Holland, pp. 151-167, 1973.

408.   King, G.L., and Buchwald, S. Characterization and partial purification of an endothelial cell growth factor from human platelets. J. Clin. Invest. 73:392-396, 1984.

409.   Kissun, R.D., Hill, C.R., Garner, A., Phillips, P., Kumar, S., and Weiss, J.B. A low-molecular-weight angiogenic factor in cat retina. Br. J. Ophthalmol. 66: 165-169, 1982.

410.   Klagsbrun, M., Knighton, D., and Folkman, J. Tumor angiogenesis activity in cells grown in tissue culture. Cancer Res. 36: 110-114, 1976.

411.   Klagsbrun, M., Sasse, J., Sullivan, R., and Smith, J.A. Human tumor cells synthesize an endothelial cell growth factor that is structurally related to basic fibroblast growth factor. Proc. Natl. Acad. Sci. USA 83:2448-2452, 1986.

412.   Klagsbrun, M., and Shing, Y. Heparin affinity of growth factors that stimulate endothelial cell proliferation. Fed. Proc. 43:521, 1984.

413.   Klagsbrun, M., and Smith, S. Purification of a cartilage-derived growth factor. J. Biol. Chem. 255:10859-10866, 1980.

414.   Klebanoff, S.J., Vadas, M.A., Harlan, J.M.,Sparks, L.H., Gamble, J.R., Agosti, J.M., and Waltersdorph, A.M. Stimulation of neutrophils by tumor necrosis factor. J. Immunol. 136:4220-4225, 1986.

415.   Klempner, M.S., Dinarello, C.A., and Gallin, J.I. Human leukocyte pyrogen induces release of specific granule contents from human neutrophils. J. Clin. Invest. 61:1330, 1978.

416.   Klintworth, G.K. The hamster cheek pouch: An experimental model of corneal vascularization. Am. J. Pathol. 73:691-710, 1973.

417.   Klintworth, G.K. The contribution of morphology to our understanding of the pathogenesis of experimentally produced corneal vascularization. Invest. Ophthalmol. Vis. Sci. 16: 281-285, 1977.

418.   Klintworth, G.K. The cornea: structure and macromolecules in health and disease. Am. J. Pathol. 89:719-808, 1977.

419.   Klintworth, G.K. Unpublished observations.

420. Klintworth, G.K., and Burger, P.C. Neovascularization of the cornea: current concepts of its pathogenesis. Int. Ophthalmol. Clin. 23:27-39, 1983.

421. Kloner, R.A., Fishbain, M.C., and Maclean, D. Effect of hyaluronidase during the early phase of acute myocardial ischemia: an ultrastructural and morphometric analysis. Am. J. Cardiol. 40: 43-49, 1977.

422. Knedler, A., and Ham, R.G. Optimized medium for clonal growth of human microvascular endothelial cells with minimal serum. In Vitro Cell. Develop. Biol. 23:481-491, 1987.

423. Knighton, D.R., Hunt, T.K., Scheuenstuhl, H., Halliday, B.J., Werb, Z., and Banda, M. Oxygen tension regulates the expression of angiogenesis factor by macrophages. Science 221: 1293-1295, 1983.

424. Knighton, D.R., Hunt, T.K., Thakral, K.K., and Goodson, W.H., III Role of platelets and fibrin in the healing sequence: an in vivo study of angiogenesis and collagen synthesis. Ann. Surg. 196:379-388, 1982.

425. Knighton, D.R., Silver, I.A., and Hunt, T.K. Regulation of wound-healing angiogenesis - effect of oxygen gradients and inspired oxygen concentration. Surgery (St. Louis) 90:262-270, 1981.

426. Koch, A.E., Polverini, J., and Leibovich, S.J.. Induction of neovascularization by activated human monocytes. J. Leukocyte Biol. 39:233-238, 1986.

427. Koch, A.S., Polverini, P.J., and Leibovich, S.J. Stimulation of neovascularization by human rheumatoid synovial tissue macrophages. Arthr. Rheum. 29:471-479, 1986.

428. Koeppe, L. Die Mikroskopie des lebenden Auges, Berlin, Julius Springer, 1920, Vol. 1.

429. Koos, R., and LeMaire, W. In Factors Regulating Ovarian Function. Greenwald, G.S. and Terranova, P.F. (Eds.). Raven, New York, 1983, pp 191-195.

430. Kreiker, A. Ueber die Entwicklung der Gefässbildung in der Hornhaut, auf Grund von Spaltlampenbeobachtungen, Magyar Orvosi Arch. 24:65, 1923 (Abstracted Zentralbl. Ges. Ophthal. 11:288, 1924.)

431. Kuettner, K.E., Croxen, R.L., Eisenstein, R., and Sorgente, N. Proteinase inhibitor activity in connective tissues. Experientia 30:595-597, 1974.

432. Kuettner, K.E., and Pauli, B.U.: Inhibition of neovascularization by a cartilage factor. In Development of the Vascular System. J. Nugent and M. O'Connor (Eds.) London, Pitman 1983, pp. 163-173.

433. Kufoy, E.A., Pakalnis, V.A., Parks, C.D., Wells, A., Yang, C.-H., and Fox, A. Keratoconjunctivitis sicca with associated secondary uveitis elicited in rats after systemic xylazine/ketamine anesthesia. Exp. Eye Res. 49:861-871, 1989.

434. Kuizenga, A.B., van Agtmaal, E.J., van Haeringen, N.J., and Kijlstra, A. Sialic acid in human tear fluid. Exp. Eye Res. 50:45-50, 1990.

435. Kulkarni, P.S., Bhattacherjee, P., Eakins, K.E., and Srinivasan, B.D. Anti-inflammatory effects of betamethasone phosphate, dexamethasone phosphate and indomethacin on rabbit ocular inflammation induced by bovine serum albumin. Curr. Eye Res. 1:43-47, 1981.

436. Kull, F.C., Jr., Brent, D.A., Parikh I., and Cuatrecacas, P.: Chemical identification of a tumor-derived angiogenic factor. Science 236:843-845, 1987.

437. Kumar, S., Shahabuddin, S., Haboubi, N., West, D., Arnold, F., Reid, H., and Carr, T. Angiogenesis factor from human myocardial infarcts. Lancet 8346:346-368, 1983.

438. Kumar, S., West, D., Daniel, M., Hancock, A., and Carr, T. Human lung tumour cell line adapted to grow in serum-free medium secretes angiogenesis factor. Int. J. Cancer 32:461, 1983.

439. Kupriianov, V.V., and Gurina, O.I. Angiogennyĭ éffekt kolkhitsina. <u>Biull. Eksp. Biol. Med.</u> 104:347-349, 1987.

440. Kurose, M. A study on blood vascular distribution in amphibian cornea. <u>Acta Soc. Ophthal. Jap.</u> 60: 621-624, 1956.

441. Lachman, L.B. Human interleukin 1: purification and properties. <u>Fed. Proc.</u> 42:2639, 1983.

442. Langer, R., Brem, H., Falterman, K., Klein, M., and Folkman, J. Isolation of a cartilage factor that inhibits tumor neovascularization. <u>Science</u> 193:70-72, 1976.

443. Langer, R., and Folkman, J. Polymers for the sustained release of proteins and other macromolecules. <u>Nature</u> 263:797-800, 1976.

444. Langer, R., and Murray, J. Angiogenesis inhibitors and their delivery systems. <u>Appl. Biochem. Biotech.</u> 8:9-24, 1983.

445. Langham, M. Utilization of oxygen by the component layers of the living cornea. <u>J. Physiol.</u> 117:461-470, 1952.

446. Langham, M. Observations on the growth of blood vessels into the cornea: Application of a new experimental technique. <u>Br. J. Ophthalmol.</u> 37:210-222, 1953.

447. Langham, M.E. The inhibition of corneal vascularization by triethylene thiophosporamide. <u>Am. J. Ophthalmol.</u> 49:1111-1117, 1960.

448. Langston, R.H., and Pavan-Langston, D. Penetrating keratoplasty for herpetic keratitis: decision-making and management. <u>Int. Ophthalmol. Clin.</u> 15: 125-140, 1975.

449. Lantz, E., and Andersson, A.: Release of fibrinolytic activators from the cornea and conjunctiva. <u>Albrecht von Graefes Arch. Klin. Exp. Ophthalmol.</u> 219:263-267, 1982.

450. Lass, J.H., Berman, M.B., Campbell, R.C., Pavan-Langston, D., and Gage, J. Treatment of experimental herpetic interstitial keratitis with medroxyprogesterone. <u>Arch. Ophthalmol.</u> 98:520-527, 1980.

451. Lavergne, G., and Colmant, I.A. Comparative study of the action of thiotepa and triamcinolone on corneal vascularization in rabbits. <u>Brit. J. Ophthalmol.</u> 48:416-422, 1964.

452. Lazar, M., Lieberman, T.W., and Leopold, I.H. Hyperbaric oxygenation and corneal neovascularization in the rabbit. <u>Am. J. Ophthalmol.</u> 66:107-110, 1968.

453. Le, J., and Vilcek, J. Tumor necrosis factor and interleukin 1: cytokines with multiple overlapping biological activities. <u>Lab. Invest.</u> 56:234-248, 1987.

454. Lee, A., and Langer, R. Shark cartilage contains inhibitors of tumor angiogenesis. <u>Science</u> 221:1185-1187, 1983.

455. Leibovich, S.J., and Polverini, P.J. Partial purification of macrophage-derived growth factor (MDGF) and macrophage-derived angiogenic activity (MDAA) by gel filtration high-pressure liquid chromatography. <u>Br. J. Rheumatol.</u> 24 (Suppl. 1) 197-202,1985.

456. Leibovich, S.J., Polverini, P.J., Shepard, H.M., Wiseman, D.M., Shively, V., and Nuseir, N. Macrophage-induced angiogenesis is mediated by tumor necrosis factor-α. <u>Nature</u> 329:630-632, 1987.

457. Leibovich, S.J., and Ross, R. The role of the macrophage in wound repair: a study with hydrocortisone and antimacrophage serum. <u>Am. J. Pathol.</u> 78:71-100, 1975.

458. Lemmon, S.K., Riley, M.C., Thomas, K.A., Hoover, G.A., Maciag, T., and Bradshaw, R.A. Bovine fibroblast growth factor: comparison of brain and pituitary preparations. <u>J. Cell Biol.</u> 95:162-169, 1982.

459. Leopold, I.H., Purnell, J.E., Cannon, E.J., Steinmetz, C.G., and McDonald, P.R. Local and systemic cortisone in ocular disease. Am. J. Ophthalmol. 34:361-371, 1951.

460. Leopold, I.H., Yeakel, E., and Calkins, L.L.: Corneal vascularization in the gray Norway rat. Arch. Ophthalmol. 42:185-187, 1949.

461. Leung, D.W., Cachianes, G., Kuang, W.-J., Goeddel, D.V., and Ferrara, N. Vascular endothelial growth factor is a secreted angiogenic mitogen. Science 246:1306-1309, 1989.

462. Leure-Dupree, A.E. Vascularization of the rat cornea after prolonged zinc deficiency. Anat. Rec. 216:27-32, 1986.

463. Levene, R., Shapiro, A., and Baum, J. Experimental corneal vascularization. Arch. Ophthalmol. 70:242-249, 1963.

464. Levin E.G., and Loskutoff, D.J. Cultured bovine endothelial cells produce both urokinase and tissue-type plasminogen activators. J. Cell Biol. 94:631-636, 1982.

465. Lewis, P.A., and Montgomery, C.M. Experimental tuberculosis of the cornea. J. Exp. Med. 20:269-281, 1914.

466. Lewkowicz-Moss, S.J., Shimeld, C., Lipworth, K., Hill, T.J., Blyth, W.A., and Easty, D.L. Quantitative studies on Langerhans cells in mouse corneal epithelium following infection with herpes simplex virus. Exp. Eye Res. 45:127-140, 1987.

467. Libby, P., Ordovas, J.M., Auger, K.R., Robbins, A.H., Birinyi, L.K., and Dinarello, C.A. Endotoxin and tumor necrosis factor induce interleukin-1 gene expression in adult human vascular endothelial cells. Am. J. Pathol. 124:179, 1986.

468. Lin, M.T., Chen, Y.L., and Lue, C.M. The involvement of collagenase from polymorphonuclear leucocyte (PMN) in PGE_1 induced corneal neovascularization. Fed. Proc. 2:A1715, 1988.

469. Lister, A., and Greaves, D.P. Effect of cortisone upon the vascularization which follows corneal burns. I. Heat burns. Brit. J. Ophthalmol. 35:725-9, 1951.

470. Liston, A.J. Corneal vascularization. Can. J. Ophthalmol. 21:30, 1986.

471. Liu, S.H., Tagawa, Y., Prendergast, R.A., Franklin, R.M., and Silverstein, A.M. Secretory component of IgA: a marker of differentiation of ocular epithelium. Invest. Ophthalmol. Vis. Sci. 20:100, 1981.

472. Lobb, R.R. Clinical applications of heparin-binding growth factors. Eur. J. Clin. Invest. 18:321-336, 1988.

473. Lobb, R., Sasse, J., Sullivan, R., Shing, Y., D'Amore, P., Jacobs, J., and Klagsbrun, M. Purification and characterization of heparin-binding endothelial cells growth factors. J. Biol. Chem. 261:1924-1928, 1986.

474. Lobb, R.R., Alderman, E.M., and Fett, J.W. Induction of angiogenesis by bovine brain derived class 1 heparin-binding growth factor. Biochemistry 24:4969-4973, 1985.

475. Lobb, R.R., and Fett, J.W. Purification of two distinct growth factors from bovine neural tissue by heparin affinity chromatography. Biochemistry 23:6295-6299, 1984.

476. Lobb, R.R., Harper, J.W., and Fett, J.W. Review: Purification of heparin-binding growth factors. Anal. Biochem. 154:1-14, 1986.

477. Loeb, J. Ueber die Entwicklung von Fischembryonen ohne Kreislauf. Arch. gesammte Physiol. Bonn. 54:525-531, 1893.

478. Long, E.R., Holley, S.W., and Vorwald, A.J. A comparison of the cellular reaction in experimental tuberculosis of the cornea in animals of varying resistance. Am. J. Pathol. 9:329-336, 1933.

479. Loskutoff, D.J., and Edgington, T.S. Synthesis of a fibrinolytic activator and inhibitor by endothelial cells. Proc. Natl. Acad. Sci. USA 74:3903-3907, 1977.

480. Lukes, R.J., and Tindle, B.H. Immunoblastic lymphadenopathy: a hyperimmune entity resembling Hodgkin's disease. N. Engl. J. Med. 292:1-8, 1975.

481. Lutty, G.A., Chandler, C., Bennett, A., Fait, C., and Patz, A. Presence of endothelial cell growth factor activity in normal and diabetic eyes. Curr. Eye Res. 5:9-16, 1986.

482. Lutty, G.A., Thompson, D.C., Gallup, J.Y., Mello, R.J., Patz, A., and Fenselau, A. Vitreous: an inhibitor of retinal extract-induced neovascularization. Invest. Ophthalmol. Vis. Sci. 24:52-56, 1983.

483. Lyons-Giordano, B., Brinker, J.M., and Kefalides, N.A. The effect of heparin on fibronectin and thrombospondin synthesis and mRNA levels in cultured human endothelial cells. Exp. Cell Res. 186:39-46, 1990.

484. Maca, R.D., Fry, G.L., Hoak, J.C., and Loh, P.T. The effects of intact platelets on cultured human endothelial cells. Thromb. Res. 11:715-727, 1977.

485. Mach, K.W., and Wilgram, G.F. Characteristic histopathology of cutaneous lymphoplasia (lymphocytoma). Arch. Dermatol. 94:26-32, 1966.

486. Maciag, T. Molecular and cellular mechanisms of angiogenesis. Important Advances in Oncology V.T. Devita, Jr., S. Hellman, S.A. Rosenberg (Eds.), Philadelphia, Lippincott, 1990, pp. 85-98.

487. Maciag, T., Cerundolo, J., Ilsley, S., Kelley, P.R., and Forand, R. An endothelial cell growth factor from bovine hypothalamus: identification and partial characterization. Proc. Natl. Acad. Sci. USA 76:5674-5678, 1979.

488. Maciag, T., Hoover, G.A., and Weinstein, R. High and low molecular weight forms of endothelial cell growth factor. J. Biol. Chem. 257:5333-5336, 1982.

489. Maciag, T., Kadish, J., Wilkins, L., Stemerman, M.B., and Weinstein, R. Organizational behavior of human umbilical vein endothelial cells. J. Cell Biol. 94:511-520, 1982.

490. Maciag, T., Mehlman, T., Friesel, R., and Schreiber, A.B. Heparin binds endothelial cell growth factor, the principal endothelial cell mitogen in bovine brain. Science 225:932-935, 1984.

491. Madri, J.A., and Pratt, B.M. In vitro models of angiogenesis. J. Histochem. Cytochem. 34:85, 1986.

492. Madri, J.A., Pratt, B.M., and Tucker, A.M. Phenotypic modulation of endothelial cells by transforming growth factor-β depends on the composition and organization of the extracellular matrix. J. Cell Biol. 106:1375-1384, 1988.

493. Madri, J.A., and Williams, S.K.: Capillary endothelial cell cultures: phenotypic modulation by matrix components. J. Cell Biol. 97:153, 1983.

494. Mahoney, J.M., and Waterbury, L.D. Drug effects on the neovascularization response to silver nitrate cauterization of the rat cornea. Curr. Eye Res. 4:531-535, 1985.

495. Maione, T.E., Gray, G.S., Petro, J., Hunt, A.J., Donner, A.L., Bauer, S.I., Carson, H.F., and Sharpe, R.J. Inhibition of angiogenesis by recombinant human platelet factor-4 and related peptides. Science 247:77-79, 1990.

496. Mann, I., Pirie, A., and Pullinger, B.D. An experimental and clinical study of the reaction of the anterior segment of the eye to chemical injury, with special reference to chemical warfare agents. Br. J. Ophthalmol. Suppl. 13:5-171, 1948.

497. Marczak, M., Majewski, S., Skopinska-Rozewska, E., Polakowski, I., and Jablonska, S. Enhanced angiogenic capability of monocyte enriched

mononuclear cell suspension from patients with systemic sclerosis. J. Invest.Dermatol. 86:355-358, 1986.

498. Marks, R.M., Roche, W.R., Czerniecki, M., Penny, R., and Nelson, D.S.: Mast cell granules cause proliferation of human microvascular endothelial cells. Lab. Invest. 55:289-294, 1986.

499. Marsh, R.J. Argon laser treatment of lipid keratopathy. Br. J. Ophthalmol. 72:900-904, 1988.

500. Martin, B.M., Gimbrone, M.A., Jr., Majeau, G.R., Unanue, E.R., and Cotran, R.S. Stimulation of human monocyte- and macrophage-derived growth factor (MDGF) production by plasma fibronectin. Am. J. Pathol. 111:367-373, 1983.

501. Martin, B.M., Gimbrone, M.A., Jr., Unanue, E.R., and Cotran, R.S. Stimulation of nonlymphoid mesenchymal cell proliferation by a macrophage-derived growth factor. J. Immunol. 126:1510-1515, 1981.

502. Matsubara, T., and Ziff, M. Increased superoxide anion release from human endothelial cells in response to cytokines. J. Immunol. 137:3295-3298, 1986.

503. Matsubara, T., and Ziff, M. Inhibition of human endothelial cell proliferation by gold compounds. J. Clin. Invest. 79:1440-1446, 1987.

504. Matsuda, H., and Sugiura, S. Ultrastructure of "tubular body" in the endothelial cells of the ocular blood vessels. Invest. Ophthalmol. 9: 919-925, 1970.

505. Matsuhashi, K. Experimental study of the corneal vascularization. Acta Soc. Ophthalmol. Jap. 66:939, 1962 (abstracted Ophthal. Lit. London 16:3992).

506. Maumenee, A.E. Discussion, The Transparency of the Cornea. Edited by S. Duke-Elder, E.S. Perkins. Oxford, Blackwell Scientific Publications, 1960.

507. Maumenee, A.E., and Kornblueth, W. Regeneration of the corneal stromal cells: Review of literature and histologic study. Am. J. Ophthalmol. 32:1051-1064, 1949.

508. Maumenee, A.E., and Scholz, R.O. The histopathology of the ocular lesions produced by the sulfur and nitrogen mustards. Bull. Johns Hopkins Hosp. 82:121,1948.

509. Maun, M.E., Cahill, W.M., and Davis, R.M. Morphologic studies of rats deprived of essential amino acids. III. Histidine. Arch. Pathol. 41:25-31, 1946.

510. Maurice, In "The Eye" (H. Davson, ed.)., Academic Press, New York, Vol. 1, 2nd Ed., p. 489, 1969.

511. Maurice, D., and Perlman, M. Permanent destruction of the corneal endothelium in rabbits. Invest. Ophthalmol. Vis. Sci. 16: 646-649, 1977.

512. Maurice, D.M., Zauberman, H., and Michaelson, I.C. The stimulus to neovascularization in the cornea. Exp. Eye Res. 5:168-184, 1966.

513. McAuslan, B.R. A new theory of neovascularization based on identification of an angiogenic factor and its effect on cultured endothelial cells. In EMBO Symposium on Specific Growth Factors, Rome 9-11, Raven Press, New York, 1980.

514. McAuslan, B.R., Bender, V., Riley, W., and Moss, B.A.: New functions of epidermal growth factor: stimulation of capillary endothelial cell migration and matrix dependent proliferation. Cell Biol. Inst. Rep. 9:175-182, 1985.

515. McAuslan, B.R., and Gole, G.A.: Cellular and molecular mechanisms in angiogenesis. Trans. Ophthalmol. Soc. UK 100:354-358, 1980.

516. McAuslan, B.R., Hannan, G.N., and Reilly, W. Characterization of an endothelial cell proliferation factor from cultured 3T3 cells. Exp. Cell Res. 128:95-101, 1980.

517. McAuslan, B.R., Hannah, G.N., Reilly, W., and Stewart, F.H.C. Variant endothelial cells. Fibronectin as a transducer of signals for migration and neovascularization. J. Cell. Physiol. 104:177-186, 1980.

518. McAuslan, B.R., and Hoffman, H. Endothelium stimulating factor from Walker carcinoma - relation to tumour angiogenic factor. Exp. Cell Res. 119:181-190, 1979.

519. McAuslan, B.R., and Reilly, W. Endothelial cell phagokinesis in response to specific metal ions. Exp. Cell Res. 130:147-157, 1980.

520. McAuslan, B.R., and Reilly, W. Selenium-induced cell migration and proliferation: relevance to angiogenesis and microangiopathy. Microvasc. Res. 32:112-120, 1986.

521. McAuslan, B.R., Reilly, W., and Hannan, G.N.: Inducers of neovascularization: criteria for definition of a putative direct acting angiogenic factor. In: Progress in Microcirculation Research II, Courtice F.C., Garlick, D.G., and Perry M.A. (Eds.), Proceedings of Second Australia and New Zealand Symposium on the Microcirculation, Sydney, New South Wales, 1983.

522. McAuslan, B.R., Reilly, W.G., Hannan, G.N., and Gole, G.A. Angiogenic factors and their assay: activity of formyl methionyl leucyl phenylalanine, adenosine diphosphate, heparin, copper, and bovine endothelium stimulating factor. Microvasc. Res. 26:323-338, 1983.

523. McCarroll, D.R., Levin, E.G., and Montgomery, R.R. Endothelial cell synthesis on von Willebrand antigen II Von Willebrand factor and Von Willebrand factor/Von Willebrand antigen II complex. J. Clin. Invest. 75:1089-1095, 1985.

524. McCracken, J.S., Burger, P.C., and Klintworth, G.K. Morphologic observations on experimental corneal vascularization in the rat. Lab. Invest. 41:519-530, 1979.

525. McCracken, J.S., and Klintworth, G.K. Ultrastructural observations on experimentally produced melanin pigmentation of the corneal epithelium. Am. J. Pathol. 85:167-182, 1976.

526. Mead, A.W. Vascularity in the reptile spectacle. Invest. Ophthalmol. 15:587-591, 1976.

527. Mello, R., Fenselau, A., Lutty, G., Bennett, A., and Patz, A. Inhibition by vitreous of in vitro endothelial cell growth: apparent activation by heat. Invest. Ophthalmol. Vis. Sci. 20 (Suppl):236, 1981.

528. Mello, R.J., Lutty, G., Giles, L., Bennett, A., and Schroeder, D. Partial characterization of a vitreous derived inhibitor of vascular endothelial cell proliferation. Invest. Ophthalmol. Vis. Sci. 22(Suppl):167, 1982.

529. Mendelsohn, A.D., Stock, E.L., Lo, G.G., and Schneck, G.L. Laser photocoagulation of feeder vessels in lipid keratopathy. Ophthal. Surg. 17:502-508, 1986.

530. Mendelsohn, A.D., Watson, Alfonso, E.C., Lieb, M., Mendelsohn, G.P., Forster, R.K., and Dennis, J.J. Amelioration of experimental lipid keratopathy by photochemically induced thrombosis of feeder vessels. Arch. Ophthalmol. 105:983-988, 1987.

531. Mergia, A., Eddy, R., Abraham, J.A., Fiddes, J.C., and Shows, T.B. The genes for basic and acidic fibroblast growth factor are on different human chromosomes. Biochem. Biophys. Res. Comm. 138:644-651, 1986.

532. Meyer, K., and Chaffee, E. The mucopolysaccharide acid of the cornea and its enzymatic hydrolysis. Am. J. Ophthalmol. 23:1320-1325, 1940.

533. Meyer, R.F., Smolin, G., Hall, J.M., and Okumoto, M. Effect of local corticosteroids on antibody-forming cells in the eye and draining lymph nodes. Invest Ophthalmol. 14:138-144, 1975.

534. Michaelson, I.C. The mode of development of the vascular system of the retina with some observations on its significance for certain retinal diseases. Trans. Ophthalmol. Soc. UK 68:137, 1948.

535. Michaelson, I.C. Effect of cortisone upon corneal vascularization produced experimentally. Arch. Ophthalmol. 47:459-464, 1952.

536. Michaelson, I.C., Herz, N., and Kertecz, D. Effect of increased oxygen concentration on new vessel growth in the adult cornea. Brit. J. Ophthalmol. 38:588-590, 1954.

537. Michaelson, I.C., Herz, N., and Rapoport, G. Effect of hyaluronidase on new vessel formation in the cornea: an experimental study. Arch. Ophthalmol. 50:613-617, 1953.

538. Middleton, D.G., and McCulloch, C. An enquiry into characteristics of sutures, particularly fine sutures. Bibl. Ophthal. 81:35-48, 1970.

539. Miki, H., Yamane, A., Tokura, T., and Sano, T. Corneal neovascularization after anterior uveal ischemia by occlusion of both long ciliary arteries in rabbit's eye. Proc. Internat. Soc. Eye Res. 4:17, 1986.

540. Miles, L.A., and Plow, E.F. Binding and activation of plasminogen on the platelet surface. J. Biol. Chem. 260:4303, 1985 (to cite).

541. Modat, G., Muller, A., Mary A., Grégoire, C., and Bonne, C. Differential effects of leukotrienes B_4 and C_4 on bovine aortic endothelial cell proliferation in vitro. Prostaglandins 33:531-538, 1987.

542. Montesano, R., Mossaz, A., Ryser, J.-E., Orci, L., and Vasalli, P. Leukocyte interleukins induce cultured endothelial cells to produce a highly organized, glycosaminoglycan-rich pericellular matrix. J. Cell Biol. 99:1706-1715, 1984.

543. Montesano, R., and Orci, L. Tumor-promoting phorbol esters induce angiogenesis in vitro. Cell 42:469-477, 1985.

544. Montesano, R., and Orci, L. Phorbol esters induce angiogenesis in vitro from large-vessel endothelial cells. J. Cell Physiol. 130:284-291, 1987.

545. Montesano, R., Orci, L., and Vassalli, P. In vitro rapid organization of endothelial cells into capillary-like networks is promoted by collagen matrices. J. Cell Biol. 97:1648, 1983.

546. Montesano, R., Pepper, M.S., Vassalli, J.D., and Orci, L. Phorbol ester induces cultured endothelial cells to invade a fibrin matrix in the presence of fibrinolytic inhibitors. J. Cell. Physiol. 132:509-516, 1987.

547. Montesano, R., Vassalli, J.D., Baird, A., Guillemin, R., and Orci, L. Basic fibroblast growth factor induces angiogenesis in vitro Proc. Natl. Acad. Sci. 83:7297-7301, 1986.

548. Moolenaar, W.H., Tertoolen, L.G.J., and de Laat, S.W. Na^+/H^+ exchange and cytoplasmic pH in the action of growth factors in human fibroblasts. Nature (London) 304:645-648, 1983.

549. Moore, F., and Riordan, J.F. Angiogenin activates phospholipase C and elicits a rapid incorporation of fatty acid into cholesterol esters in vascular smooth muscle cells. Biochemistry 29:228-233, 1990.

550. Moore, J.W., and Sholley, M.M. Comparison of the neovascular effects of stimulated macrophages and neutrophils in autologous rabbit corneas. Am. J. Path. 120: 87-98, 1985.

551. Morris, P.B., Hida, T., Blackshear, P.J., Klintworth, G.K., and Swain, J.L. Tumor-promoting phorbol esters induce angiogenesis in vivo. Am. J. Physiol. 254:(Cell Physiol.) C318-322, 1988.

552. Moscat, J., Moreno, F., Herrero, C., Lopez, C., and Garcia-Barreno, P. Endothelial cell growth factor and ionophore A23187 stimulation of production of inositol phosphates in porcine aortic endothelial cells. Proc. Natl. Acad. Sci. USA 85:659-663, 1988.

553. Moscatelli, D., Presta, M., and Rifkin, D.B. Purification of a factor from human placenta that stimulates capillary endothelial cell protein production,

DNA synthesis and migration. <u>Proc. Natl. Acad. Sci. USA</u> 83:2091-2095, 1986.

554. Moses, M.A., Sudhalter, J., and Langer, R. Identification of an inhibitor of neovascularization from cartilage. <u>Science</u> 248:1408-1410, 1990.

555. Mosesson, M., and Doolittle, F. The conversion of fibrinogen to fibrin: events and recollections from 1942 to 1982. <u>Ann NY Acad. Sci.</u> 408:1-672, 1983.

556. Mostafa, L.K., Jones, D.B., and Wright, D.H. Mechanism of the induction of angiogenesis by human neoplastic lymphoid tissue: studies employing bovine aortic endothelial cells <i>in vitro</i>. <u>J. Pathol.</u> 132:207-216, 1980.

557. Moticka, E.J., and She, S.-C. Comparison of failure rates of orthotopic corneal grafts using three different grafting procedures. <u>Curr. Eye Res.</u> 8:813-820, 1989.

558. Motro, B., Itin, A., Sachs, L., and Keshet, E. Pattern of interleukin 6 gene expression <i>in vivo</i> suggests a role for this cytokine in angiogenesis. <u>Proc. Natl. Acad. Sci. USA</u> 87:3092-3096, 1990.

559. Movat, H.Z. <u>The Inflammatory Reaction</u>. Amsterdam, Elsevier Biochemical Publishers, 1985, pp. 1-365.

560. Muller, G., Behrens, J., Nussbaumer, U., Bohlen, P., and Birchmeier, W. Inhibitory action of transforming growth factor β on endothelial cells. <u>Proc. Natl. Acad. Sci. USA</u> 84:5600-5604, 1987.

561. Mulliken, J.B., Zetter, B.R., and Folkman, J. In vitro characteristics of endothelium from hemangiomas and vascular malformations. <u>Surgery</u> 1982, 92:348-353.

562. Mullins, D.E., and Rifkin, D.B. Stimulation of motility in cultured bovine capillary endothelial cells by angiogenic preparations. <u>J. Cell. Physiol.</u> 119: 247-254, 1984.

563. Muthukkaruppan, VR., and Auerbach, R. Angiogenesis in the mouse cornea. <u>Science</u> 205:1416-1418, 1979.

564. Muthukkaruppan, V.R., Kubai, L., and Auerbach, R. Tumor-induced neovascularization in the mouse eye. <u>J. Nat. Cancer Inst.</u> 69:699-708, 1982.

565. Nakayasu, K. Origin of pericytes in neovascularization of rat cornea. <u>Jap. J. Ophthalmol.</u> 32:105-112, 1988.

566. Nathan, C.F. Secretory products of macrophages. <u>J. Clin. Invest.</u> 79:319, 1987.

567. Nawroth, P., Bank, I., Handley, D., Cassimeris, J., Chess, L., and Stern, D. Tumor necrosis factor/cachectin interacts with endothelial cell receptors to induce release of interleukin-1. <u>J. Exp. Med.</u> 163:1363, 1986.

568. Nesburn, A. Complications associated with therapeutic soft contact lenses. <u>Ophthalmology</u> 86:1130-1137, 1979.

569. Nicol, J.A.C. <u>The Eyes of Fishes</u>, New York, Oxford University Press, pp. 31-33,1989.

570. Nicosia, R.F., Tchao, R., and Leighton, J. Histotypic angiogenesis in vitro: light microscopic, ultrastructural, and radioautographic studies. <u>In Vitro</u> 18:538-549, 1982.

571. Niederkorn, J.Y., Ubelaker, J.E., and Martin, J.M. Vascularization of corneas of hairless mutant mice. <u>Invest. Ophthalmol. Vis. Sci.</u> 31:948-953, 1990.

572. Nikolic, L., Friend, J., Taylor, S., and Thoft, R.A. Inhibition of vascularization in rabbit corneas by heparin:cortisone pellets. <u>Invest. Ophthalmol. Vis. Sci.</u> 27:449-456, 1986.

573. Nirankari, V.S., and Baer, J.C. Corneal argon laser photocoagulation for neovascularization in penetrating keratoplasty. <u>Ophthalmology</u> 93:1304-1309, 1986.

574. Nishizuka, Y. The role of protein kinase C in cell surface signal transduction and tumour promotion. Nature 308:693-698, 1984.

575. Noji, S., Matsuo, T., Koyama, E., Yamaai, T., Nohno, T., Matsuo, N., and Taniguchi, S. Expression pattern of acidic and basic fibroblast growth factor genes in adult rat eyes. Biochem. Biophys. Res. Commun. 168:343-349, 1990.

576. Norman, M.G., and O'Kusky, J.R. The growth and development of microvasculature in human cerebral cortex. J. Neuropath. Exp. Neurol. 45: 222-232, 1986.

577. Norrby, K., Jakobsson, A., and Sörbo, J. Mast-cell secretion and angiogenesis, a quantitative study in rats and mice. Virchows Archiv. B. Cell Pathol. 57:251-256, 1989.

578. Norris, D.A., Clark, R.A.F., Swigart, L.M., Huff, J.C., Weston, W.L., and Howell, S.E. Fibronectin fragments are chemotactic for human peripheral blood monocytes. J. Immunol. 129:1612-1618, 1982.

579. Nunziata, B., Smith, R.S., and Weimar, V. Corneal radiofrequency burns: effects of prostaglandins and 48/80. Invest. Ophthalmol. Vis. Sci. 16: 285-291, 1977.

580. Obenberger, J. Calcification in corneas with alloxan-induced vascularization. Am. J. Ophthalmol. 68:113-1119, 1969.

581. O'Donoghue, M.N., and Zarem, H.A. Stimulation of neovascularization-comparative efficacy of fresh and preserved skin grafts. Plastic Reconstr. Surg. 48:474-477, 1971.

582. Oehlschlaegel, G., Stollmann, K., and Schröpl, F. Ungewöhnliche Hämangiomatose der Haut bei Plasmocytose. Hautarzt 19:210-215, 1968.

583. Ohhabra, H.N., and Consul, B.N. Oxygen in corneal vascularization. J. All India Ophthalmol. Soc. 18: 41-44, 1970.

584. Ohhabra, H.N., and Sharma, D.P. Subconjunctival oxygen in corneal vascularization. Eye Ear Nose Throat Mon. 50: 17-19, 1971.

585. Old, L.J. Tumor necrosis factor (TNF). Science 230:630-632, 1985.

586. Olson, C.L. Subconjunctival steroids and corneal hypersensitivity. Arch. Ophthalmol. 75:651-658, 1966.

587. Ooi, B.S., MacCarthy, E.P., Hsu, A., and Ooi, Y.M. Human mononuclear cell modulation of endothelial cell proliferation. J. Lab. Clin. Med. 102:428-434, 1983.

588. Oppenheim, J.J., and Gery, I. Interleukin 1 is more than an interleukin. Immunol. Today 3:113-119, 1982.

589. Oppenheim, J.J., Kovacs, E.J., Matsushima, K., and Durum, S.K. There is more than one interleukin 1. Immunol. Today 7:46, 1986.

590. Ormerod, L.D., Abelson, M.B., and Kenyon, K.R. Standard models of corneal injury using alkali-immersed filter discs. Invest. Ophthalmol. Vis. Sci. 30:2148-2153, 1989.

591. Pandolfi, M., and Astrup, T. A histochemical study of the fibrinolytic activity: cornea, conjunctiva, and lacrimal gland. Arch. Ophthalmol. 77:258-264, 1967.

592. Paque, J., and Poirer, R.H. Corneal allograft reaction and its relationship to suture site neovascularization. Ophthalmic Surg. 8: 71-74, 1977.

593. Parke A., Bhattacherjee, P., Palmer, R.M.J., and Lazarus, N.R. Characterization and quantification of copper sulfate-induced vascularization of the rabbit cornea. Am. J. Path. 130:173-178, 1988.

594. Parker, P.J., Coussens, L., Totty, N., Rhee, L., Young, S., Chen, E., Stabel, S., Waterfield, M.D., and Ullrich, A. The complete primary structure of protein kinase C - the major phorbol ester receptor. Science 233:853-866, 1986.

595. Pepper, M.S., Vassalli, J.D., Montesano, R., and Orci, L. Urokinase-type plasminogen activator is induced in migrating capillary endothelial cells. J. Cell Biol. 105:2535-2541, 1987.

596. Peterson, H-I. Tumor angiogenesis inhibition by prostaglandin synthetase inhibitors. Anticancer Research 6:251-254, 1986.

597. Phillips, K., Arffa, R., Cintron, C., Rose, J., Miller, D., Kublin, C.L., and Kenyon, K.R. Effects of prednisolone and medroxyprogesterone on corneal wound healing, ulceration, and neovascularization. Arch. Ophthalmol. 101:640-643, 1983.

598. Phillips, P., and Kumar, S. Tumour angiogenesis factor (TAF) and its neutralisation of a xenogenic antiserum. Int. J. Cancer 23: 82-88, 1979.

599. Phillips, P., Steward, J.K., and Kumar, S. Tumor angiogenesis factor (TAF) in human and animal tumors. Int.J. Cancer 17:549-558, 1976.

600. Piguet, P.F., Grau, G.E., and Vassalli, P. Subcutaneous perfusion of tumor necrosis factor induces local proliferation of fibroblasts, capillaries, and epidermal cells, or massive tissue necrosis. Am. J. Path. 136:103-110, 1990.

601. Plouët, J., Schilling, T., and Gospodarowicz, D. Isolation and characterization of a newly identified endothelial cell mitogen produced by AtT-20 cells. EMBO J 8:3801-3806, 1989.

602. Plunkett, M.L., and Hailey, J.A. Methods in Laboratory Investigation. An in vivo quantitative angiogenesis model using tumor cells entrapped in alginate. Lab. Invest. 62:510-517, 1990.

603. Pober, J.S., Bevilacqua M.P., Mendick, D.L., Lapierre, L.A., Fiers, W., and Gimbrone, M.A., Jr. Two-distinct monokines, interleukin-I and tumor necrosis factor, each independently induce biosynthesis and transient expression of the same antigen on the surface of cultured human vascular endothelial cells. J. Immunol. 136:1680-1687, 1986.

604. Pober, J.S., and Cotran, R.S. Cytokines and endothelial cell biology. Physiol. Reviews 70:427-451, 1990.

605. Pober, J.S., Gimbrone, M.A., Jr., Collins, T., Cotran, R.S., Ault, K.A., Fiers, W., Krensky, A.M., Clayberger, C., Reiss, C.S., and Burakoff, S.J. Interactions of T lymphocytes with human vascular endothelial cells: role of endothelial cells surface antigens. Immunobiol. 168:483-494, 1984.

606. Pober, J.S., Gimbrone, M.A., Jr., Cotran, R.S., Reiss, C.S., Burakoff, S.J., Fiers, W., and Ault, K.A. Ia expression by vascular endothelium is inducible by activated T cells and by human γ interferon. J. Exp. Med. 157:1339-1353, 1983.

607. Polverini, P.J., Cotran, R.S., Gimbrone, M.A., Jr., and Unanue, E.R. Activated macrophages induce vascular proliferation. Nature 269:804-806, 1977.

608. Polverini, P.J., Cotran, R.S., and Sholley, M.M. Endothelial proliferation in the delayed hypersensitivity reaction: an autoradiographic study. J. Immunol. 118:529-532, 1977.

609. Polverini, P.J., and Leibovich, S.J. Induction of neovascularization in vivo and endothelial proliferation in vitro by tumor-associated macrophages. Lab. Invest. 51:635-642, 1984.

610. Polverini, P.J., and Leibovich, S.J. Induction of neovascularization and non-lymphoid mesenchymal cell proliferation by macrophage cell lines. J. Leukocyte Biol. 37:279-288, 1985.

611. Preis, I., Langer, R., Brem, H., Folkman, J., and Patz, A. Inhibition of neovascularization by an extract derived from vitreous. Am. J. Ophthalmol. 84:323-328, 1977.

612. Proia, A.D., Chandler, D.B., Haynes, W.L., Smith, C.F., Suvarnamani, C., Erkel, F.H., and Klintworth, G.K. Methods in Laboratory Investigation.

Quantitation of corneal neovascularization using computerized image analysis. Lab. Invest. 58:473-479, 1988.

613. Proia, A.D., Chung, S.M., Klintworth, G.K., and Lapetina, E.G. Cholinergic stimulation of phosphatidic formation by rat cornea in vitro. Invest. Ophthalmol. Vis. Sci. 27:905-908, 1986.

614. Proia, A.D., Raskauskas, P.A., Hida, T., and Klintworth, G.K. Induction of corneal angiogenesis by constituents of omentum. Invest. Ophthalmol. Vis. Sci. 27(Suppl):199, 1986.

615. Radnot, M., and Jobbagyi, P. Vascularization of the cornea in ischemia of the anterior eye segment. Klin. Monatsbl. Augenheilkd. 153: 840-844, 1968.

616. Raju, K.S. Isolation and characterization of copper-binding sites of human ceruloplasmin. Mol. Cell. Biochem. 56:81-88, 1983.

617. Raju, K.S., Alessandri, G., and Gullino, P.M. Characterization of a chemoattractant for endothelium induced by angiogenesis effectors. Cancer Res. 44:1579-1584, 1984.

618. Raju, K.S., Alessandri, G., Ziche, M., and Gullino, P.M. Ceruloplasmin, copper ions, and angiogenesis. J. Natl. Cancer Inst. 69:1183-1188, 1982.

619. Rappaport, H., and Moran, E.M. Angio-immunoblastic (immunoblastic) lymphadenopathy. N. Engl. J. Med. 292:42-43, 1975.

620. Redmer, D.A., Kirsch, J.D., and Grazul, A.T. In vitro production of angiotropic factor by bovine corpus luteum: partial characterization of activities that are chemotactic and mitogenic for endothelial cells. In Regulation of Ovarian and Testicular Function. Eds. E. Anderson, D. Dhindsa, S. Kalra and V. Mahesh, Plenum Press, New York, In Press 1987.

621. Reed, J.W., Fromer, C.H., and Klintworth, G.K. Induced corneal vascularization remission with argon laser therapy. Arch. Ophthalmol. 93:1017-1019, 1975.

622. Reiko, T., and Werb, Z. Secretory products of macrophages and their physiological functions. Am. J. Physiol. 246:C1-C9, 1984.

623. Reilly, T.M., Taylor, D.S., Herblin, W.F., Thoolen, M.J., Chiu, A.T., Watson, D.W., and Timmermans, P.B.M.W.M. Monoclonal antibodies against basic fibroblast growth factor which inhibit its biological activity in vitro and in vivo. Biochem. Biophys. Res. Commun. 164:736-743, 1989.

624. Reim, M., Kaufhold, F.M., Kehrer, Th, Kuckelkorn, R., Kuwert, T., and Leber, M. Zur Differenzierung der Therapie bei Verätzungen. Fortschr. Ophthalmol. 81:583-587, 1984.

625. Reinhold, H.S., Blachiewicz, B., and Berg-Blok, A. Reoxygenation of tumours in "sandwich" chambers. Eur. J. Cancer 15:481-489, 1979.

626. Reinhold, H.S., and van den Berg-Blok, A. Enhancement of thermal damage to the microcirculation of "sandwich" tumours by additional treatment. Eur. J. Cancer Clin. Oncol. 17:781-795, 1981.

627. Remensnyder, J.P., and Majno, G. Oxygen gradients in healing wounds. Am. J. Pathol. 52:301-323, 1968.

628. Reynolds, L.P., Millaway, D.S., Kirsch, J.D., Infeld, J.E., and Redmer, D.A. Angiogenic activity of placental tissues of cows. J. Reprod. Fert. 81:233-240, 1987.

629. Rich, A.R., and Follis, R.H., Jr. Studies on the site of sensitivity in the Arthus phenomenon. Bull. Johns Hopkins Hosp. 66:106-122, 1940.

630. Rifkin, D.B., Gross, J.L., Moscatelli, D., and Jaffe, E. Proteases and angiogenesis: production of plasminogen activator and collagenase by endothelial cells. In "Pathobiology of the Endothelial Cell" (H.L. Nossel and H.J. Vogel, Eds). p. 191, Academic Press, New York, 1982.

631. Rifkin, D.B., and Moscatelli, D. Mini-review: recent developments in cell biology of basic fibroblast growth factor. J. Cell Biol. 109:1-6, 1989.

632. Risau, W., and Ekblom, P. Production of a heparin-binding angiogenesis factor by the embryonic kidney. J. Cell Biol. 103:1101-1107, 1986.

633. Roberts, A.B., Flanders, K.C., Kondaiah, K., Thompson, N.L., van Obberghen-Schilling, E., Wakefield, L., Rossi, P., de Crombrugghe, B., Heine, U., and Sporn, M.B. Transforming growth factor β: biochemistry and roles in embryogenesis, tissue repair and remodeling and carcinogenesis. Rec. Prog. Hormone Res. 44:157-197, 1988

634. Roberts, A.B., and Sporn, M.B. Regulation of endothelial cell growth, architecture, and matrix synthesis by TGF-β. Am. J. Respir. Dis. 140:1126-1128, 1989.

635. Roberts, A.B., Sporn, M.B., Assoian, R.K., Smith, J.M., Roche, N.S., Wakefield, L.M., Heine, U.I., Liotta, L.A., Falanga, V., Kehrl, J.H., and Fauci, A.S. Transforming growth factor type β: rapid induction of fibrosis and angiogenesis in vivo and stimulation of collagen formation in vitro. Proc. Natl. Acad. Sci. USA 83:4167-4171, 1986.

636. Robin, J.B., Regis-Pacheco, L.F., Kash, R.L., and Schanzlin, D.J. The histopathology of corneal neovascularization: inhibitor effects. Arch. Ophthalmol. 103:284-287, 1985.

637. Rochels, R. Tierexperimentelle Untersuchungen zur Rolle von Entzündungsmediatoren bei der Hornhautneovaskurisation. Docum. Ophthalmol. 57:215-262, 1984.

638. Roizenblatt, J. Interstitial keratitis caused by American (mucocutaneous) leishmaniasis. Am. J. Ophthalmol. 87:175-179, 1979.

639. Rollins, B.J., Yoshimura, T., Leonard, E.J., and Pober, J.S. Cytokine-activated human endothelial cells synthesize and secrete a monocyte chemoattractant, MCP-1/JE. Am. J. Path. 136:1229-1233, 1990.

640. Rootman, J., Bussanich, N., Gudauskas, G., and Kumi, C. Effects of subconjunctivally injected antineoplastic agents on three models of corneal inflammation. Can. J. Ophthalmol. 20:142-146, 1985.

641. Rosenbaum, J.T., Howes, E.L., Jr., Rubin, R.M., and Samples, J.R. Ocular inflammatory effects of intravitreally-injected tumor necrosis factor. Am. J. Path. 133:47-53, 1988.

642. Rosenbruch, M. Granulation tissue in the chick embryo yolk sac blood vessel system. J. Comp. Path. 101:363-373, 1989.

643. Ross R. The pathogenesis of atherosclerosis - an update. New Engl. J. Med. 314:488-500, 1986.

644. Roth, S.I., Stock, E.L., Siel, J.M., Mendelsohn, A., Reddy, C., Preskill, D.G., and Ghosh, S. Pathogenesis of experimental lipid keratopathy. Invest. Ophthal. Vis. Sci. 29:1544-1551, 1988.

645. Rozenman, Y., Donnenfeld, E.D., Cohen, E.J., Arentsen, J.J., Bernardino, V., Jr., and Laibson, P.R. Contact lens-related deep stromal neovascularization. Am. J. Ophthalmol. 107:27-32, 1989.

646. Rubsamen, P.E., McCulley, J., Bergstresser, P.R., and Streilein, J.W. On the Ia immunogenicity of mouse corneal allografts infiltrated with Langerhans cells. Invest. Ophthalmol. Vis. Sci. 25:513-518, 1984.

647. Russell, W., Proia, A.D., and Klintworth, G.K. Retrobulbar anesthesia and eyelid closure - effect on corneal angiogenesis. Cornea In Press 1990.

648. Ryan, T.J. Factors influencing the growth of vascular endothelium in the skin. Br. J. Derm. 82(Suppl 5) 99, 1970.

649. Ryan, T.J. Factors influencing growth of vascular endothelium in the skin. The Physiology and Pathophysiology of the Skin, Vol. 2. The Nerves and Blood Vessels. Edited by A. Jarrett. London, Academic Press, 1973.

650. Ryan, U.S. The endothelial surface and responses to injury. Fed. Proc. 45:101-108, 1986.

651. Ryan, U.S., and Ryan, J.W. Endothelial cells and inflammation. Clinics in Laboratory Medicine, Ward, P.A. (Ed.), Philadelphia: Saunders, 1983, pp. 577-599.

652. Ryan, U.S., and Ryan, J.W. Inflammatory mediators, contraction and endothelial cells. In Progress in Microcirculation Research, II. Courtice, F.C., Garlick, D.G., and Perry, M.A. (Eds.), Sydney, Australia: Committee in Postgraduate Medical Education, University of New South Wales, 1984, pp. 424-438.

653. Rydell, N., and Balazs, E.A. Effect of intra-articular injection of hyaluronic acid on the clinical symptoms of osteoarthritis and on granulation tissue formation. Clin. Orthop. 80: 25-32, 1971.

654. Ryu, S., and Albert, D.M. Evaluation of tumor angiogenesis factor with the rabbit cornea model. Invest. Ophthalmol. Vis. Sci. 18:831-841, 1979.

655. Saba, H.I., Hartmann, R.C., and Saba, S.R. Effect of polymorphonuclear leukocytes on endothelial cell growth. Thromb. Res. 12:397-407, 1978.

656. Saeger, F. Ueber Anregung der Gefässneubildung in der Hornhaut mit Hilfe des kaustischen Strichs, nebst Bemerkungen über die Filariaerkrankung des Auges. Klin. Mbl. Augenheilk. 82:223-229, 1929.

657. Saksela, O., Moscatelli, D., and Rifkin, D.B. The opposing effects of basic fibroblast growth factor and transforming growth factor beta on the regulation of plasminogen activator activity in capillary endothelial cells. J. Cell Biol. 105:957-963, 1987.

658. Salo, T., Liotta, L.A., Keski-Oja, J., Turpeenniemi-Hujanen, T., and Tryggvason, K. Secretion of basement membrane collagen degrading enzyme and plasminogen activator by transformed cells - role in metastasis. Int. J. Cancer 30:669-673, 1982.

659. Sandison, J.C. A new method for the microscopic study of living growing tissue by the introduction of a transparent chamber in the rabbit's ear. Anat. Rec. 28:281-287, 1924.

660. Sato, N., Fukuda, K., Nariuchi, H., and Sagara, N. Tumor necrosis factor inhibiting angiogenesis in vitro. J. Natl. Cancer Instit. 79:1383-1391, 1987

661. Sato, N., Goto, T., Haranaka, K., Satomi, N., Nariuchi, H., Mano-Hirano, Y.,and Sawasaki, Y. Actions of tumor necrosis factor on cultured vascular endothelial cells: morphologic modulation, growth inhibition, and cytotoxicity. J. Natl. Cancer Instit. 76:1113-1121, 1986.

662. Sato, N., Sawasaki, Y., Haranaka, K., Satomi, N., Nariuchi, H., and Goto, T. Growth inhibitory and cytotoxic action of rabbit tumor necrosis factor against bovine capillary endothelial cells in vitro. Proc. Jap. Acad. 61(B):471-474, 1985.

663. Sato, Y., and Rifkin, D.B. Autocrine activities of basic fibroblast growth factor: regulation of endothelial cell movement, plasminogen activator synthesis, and DNA synthesis. J. Cell Biol. 107:1199-1205, 1988.

664. Schanzlin, D.J., Cyr, R.J., and Friedlaender, M.H. The histopathology of corneal neovascularization. Arch. Ophthalmol. 101:472-474, 1983.

665. Scher, W. Biology of Disease: The role of extracellular proteases in cell proliferation and differentiation. Lab. Invest. 57:607-633, 1987.

666. Schoefl, G.I. Studies on inflammation. III. Growing capillaries: Their structure and permeability. Virchows Arch. Pathol. Anat. 337:97-141, 1963.

667. Schoefl, G.I. Electron microscopic observations on the regeneration of blood vessels after injury. Ann N.Y. Acad. Sci. 116:789-802, 1964.

668. Schor, A.M., and Schor, S.L. Tumor angiogenesis. J. Pathol. 141:385, 1983.

669. Schor, A.M., Schor, S.L., and Allen, T.D. Effects of culture conditions on the proliferation, morphology and migration of bovine aortic endothelial cells. J. Cell. Sci. 62:267-285, 1983.

670.   Schor, A.M., Schor, S.L., and Kumar, S. Importance of a collagen substratum for stimulation of capillary endothelial cell proliferation by tumour angiogenesis factor. Int. J. Cancer 24:225-234, 1979.

671.   Schor, A.M., Schor, S.L., Weiss, J.B., Brown, R.A., Kumar, S., and Phillips, P. Stimulation by a low molecular weight angiogenic factor of capillary endothelial cells in culture. Br. J. Cancer 41: 790-799, 1980.

672.   Schreiber, A.B., Kenney, J., Kowalski, W.J., Friesel, R., Mehlman, T., and Maciag, T. Interaction of endothelial cell growth factor with heparin: characterization by receptor and antibody recognition. Proc. Natl. Acad. Sci. USA 82:6138-6142, 1985.

673.   Schreiber, A.B., Kenney, J., Kowalsky, W.J., Thomas, K.A., Gimenez-Gallego, G. Rios-Canderlore, M., Di Salvo, J., Barritault, D., Courty, J., Courtois, Y., Moenner, M., Burgess, W., Mehlman, T., Friesel, T., Johnson, W., and Maciag, T. A unique family of endothelial cell polypeptide mitogens: the antigenic and receptor cross-reactivity of bovine endothelial cell growth factor, brain-derived acidic fibroblast growth factor, and eye-derived growth factor. J. Cell Biol. 101:1623-1626, 1985.

674.   Schreiber, A.B., Winkler, M.E., and Derynck, R. Transforming growth factor-α: a more potent angiogenic mediator than epidermal growth factor. Science 232:1250-1253, 1986.

675.   Schultz, D.R., and Yunis, A.A. Immunoblastic lymphadenopathy with mixed cryoglobulinemia: a detailed case study. N. Engl. J. Med. 292:8-12, 1975.

676.   Schumacher, B.L., Grant, D., and Eisenstein, R. Smooth muscle cells produce an inhibitor of endothelial cell growth. Arteriosclerosis 5:110-115, 1985.

677.   Schwartz, G. (ed). Corticosteroids and the eye. Int. Ophthalmol. Clin. 6:753-1104, 1966.

678.   Schweigerer, L., Malerstein, B., and Gospodarowicz, D. Tumor necrosis factor inhibits the proliferation of cultured capillary endothelial cells. Biochem. Biophys. Res. Commun. 143:997-1004, 1987.

679.   Schweigerer, L., Neufeld, G., Friedman, J., Abraham, J.A., Fiddes, J.C., and Gospodarowicz, D. Capillary endothelial cells express basic fibroblast growth factor, a mitogen that promotes their own growth. Nature 325:257-259, 1987.

680.   Senger, D.R., Connolly, D.T., van De Water, L., Feder, J., and Dvorak, H.F. Purification and NH_2-terminal amino acid sequence of guinea pig tumor-secreted vascular permeability factor. Cancer Res. 50:1774-1778, 1990.

681.   Senior, R.M., Huang, S.S., Griffin, G.L., and Huang, J.S. Brain-derived growth factor is a chemoattractant for fibroblasts and astroglial cells. Biochem. Biophys. Res. Commun. 141:67-72, 1986.

682.   Serafin, W.E., Katz, H.R., Austen, K.F., and Stevens, R.L. Complexes of heparin proteoglycans, chondroitin sulfate E proteoglycans, and [^{3}H] diisopropyl fluorophosphate-binding proteins are exocytosed from activated mouse bone marrow-derived mast cells. J. Biol. Chem. 261:15017-15021, 1986.

683.   Shahabuddin, S., and Kumar, S. Quantitation of angiogenesis factor in bovine retina and tumour extracts by means of radioimmunoassay. Brit. J. Ophthalmol. 67:286-291, 1983.

684.   Shahabuddin, S., Kumar, S., West, D., Arnold, F. A study of angiogenesis factors from five different sources using a radioimmunoassay. Intl. J. Cancer 35:87, 1985.

685.   Shapiro, R., Strydom, D.J., Olson, K.A., and Vallee, B.L. Isolation of angiogenesis from normal human plasma. Biochemistry 26:5141-5146, 1987.

686.   Shing, Y., Folkman, J., Haudenschild, C., Lund, D., Crum, R., and Klagsbrun, M. Angiogenesis is stimulated by a tumor-derived endothelial cell growth factor. J. Cell. Biochem. 29:275-287, 1985.

687.   Shing, Y., Folkman, J., Sullivan, R., Butterfield, C., Murray, J., and Klagsbrun, M. Heparin affinity: purification of a tumor-derived capillary endothelial cell growth factor. Science 223:1296-1299, 1984.

688.   Shio, H., and Ramwell, P. Effect of prostaglandin E_2 and aspirin on the secondary aggregation of human platelets. Nature; New Biol. 236:45-46, 1972.

689.   Sholley, M.M., Cavallo, T., and Cotran, R.S. Endothelial proliferation in inflammation: I. Autoradiographic studies following thermal injury to the skin of normal rats. Am. J. Pathol. 89:277-290, 1977.

690.   Sholley, M.M., and Cotran, R.S. Endothelial DNA synthesis in the microvasculature of rat skin during the hair growth cycle. Amer. J. Anat. 147: 243-249, 1976.

691.   Sholley, M.M., and Cotran, R.S. Endothelial proliferation in inflammation. II. Autoradiographic studies in x-irradiated leukopenic rats after thermal injury of the skin. Am. J. Pathol. 91:229-242, 1978.

692.   Sholley, M.M., Ferguson, G.P., Seibel, H.R., Montour, J.L., and Wilson, J.D.: Mechanisms of neovascularization: vascular sprouting can occur without proliferation of endothelial cells. Lab. Invest. 51:624-634, 1984.

693.   Sholley, M.M., Gimbrone, M.A., Jr., and Cotran, R.S. Cellular migration and replication in endothelial regeneration. A study using irradiated endothelial cultures. Lab. Invest. 36:18-25, 1977.

694.   Sholley, M.M., Gimbrone, M.A., Jr., and Cotran, R.S. The effects of leukocyte depletion on corneal neovascularization. Lab. Invest. 38:32-40, 1978.

695.   Sholley, M.M., Wilson, J.D., Montour, J.L., and Ruffolo, J.J., Jr. Radiation response of corneal neovascularization. ARVO Abstract. Invest. Ophthalmol. Vis. Sci. 19(Suppl.):254, 1980.

696.   Sidky, Y.A., and Auerbach, R. Lymphocyte-induced angiogenesis: a quantitative and sensitive assay of the graft-versus-host reaction. J. Exp. Med. 141: 1084- 1100, 1975.

697.   Sidky, Y.A., and Auerbach, R. Lymphocyte-induced angiogenesis in tumor-bearing mice. Science 192:1237-1238, 1976.

698.   Silva, J., Eslava, C., Basterrica, C., Steel, C., and Courtin, L. Experimental corneal neovascularization treated with beta radiation. Arch. Chil. Oftal. 26:70, 1969.

699.   Silver, I.A. The measurement of oxygen tension in healing tissue. Prog. Respir. Res. 3:124-135, 1969.

700.   Silverman, L.J., Lund, D.P., Zetter, B.R., Lainey, L.L., Shahood, J.A., Freiman, D.G., Folkman, J., and Barger, A.C. Angiogenic activity of adipose tissue. Biochem. Biophys. Res. Commun. 153:347-352, 1988.

701.   Simionescu, N., Simionescu, M., and Palade, G.E. Differentiated microdomains on the luminal surface of the capillary endothelium. I. Preferential distribution of anionic sites. J. Cell Biol. 90: 605-613, 1981.

702.   Simpson, J.G., Fraser, R.A., and Thompson, W.D. Angiogenesis and angiogenesis factors. In Structure and Function of The Endothelial Cells (Eds. Messmer, K., and Hammersen, F.) Karger, Basel (Proc. 1st Bodensee Symposium on Microcirculation), 1983.

703.   Sjaarda, R.N., Glatt, H.J., and Klintworth, G.K. Enhanced corneal neovascularization in the athymic (nude) mouse. Invest. Ophthalmol. Vis. Sci. 26(Suppl.):328, 1985.

704. Skutelsky, E., and Danon, D. Redistribution of surface anionic sites on the luminal front of blood vessel endothelium after interaction with polycationic ligand. J. Cell Biol. 71: 232-241, 1976.

705. Smelser, G.K. Discussion of Ashton, N. Corneal vascularization. In The Transparency of the Cornea. (Eds. Duke-Elder, S. and Perkins, E.S.) Oxford, Blackwell Scientific Publications, Ltd. 1960, pp. 145-147.

706. Smelser, G.K., and Ozanics, V. The effect of vascularization on the metabolism of the sulfated mucopolysaccharides and swelling properties of the cornea. Am. J. Ophthalmol. 48:418-426, 1959.

707. Smith, R.S. The development of mast cells in the vascularized cornea. Arch. Ophthalmol. 66:383-390, 1961.

708. Smith, R.S., and Smith, L.A. Effects of BP961 on corneal wound healing. ARVO Abstract. Invest. Ophthalmol. Vis. Sci. 19(Suppl.):254, 1980.

709. Smith, S.S., and Basu, P.K. Mast cells in corneal immune reaction. Canad. J. Ophthalmol. 5:175-183, 1970.

710. Snyder, D.S., and Unanue, E.R.: Corticosteroids inhibit murine macrophage Ia expression and interleukin 1 production. J. Immunol. 129:1803-1805, 1982.

711. Sommer, A., and Rifkin, D.B. Interaction of heparin with human basic fibroblast growth factor: protection of the angiogenic protein from proteolytic degradation by a glycosaminoglycan. J. Cell. Physiol. 138:215-220, 1989.

712. Sorgente, N., and Dorey, C.K. Inhibition of endothelial cell growth by a factor from cartilage. Exp. Cell Res. 128:63-71, 1980.

713. Soubrane, G., Jerdan, J., Karpouzasl, Fayein, N.A., Glaser, B., Coscas, G., Courtois, Y., and Jeanny, J.C. Binding of basic fibroblast growth factor to normal and neovascularized rabbit cornea. Invest. Ophthalmol. Vis. Sci. 31:323-333, 1990.

714. Spanel-Borowski, K. Vascularization of ovaries from golden hamsters following implantation into the chick chorioallantoic membrane. Expl. Cell Biol. 57:219-227, 1989.

715. Sporn, M.B., Roberts, A.B., Shull, J.H., Smith, J.M., and Ward, J.M. Polypeptide transforming growth factors isolated from bovine sources and used for wound healing in vivo. Science 219:1329-1331, 1983.

716. Srinivasan, B.D. Corneal reepithelization and anti-inflammatory agents. Tr. Am. Ophthalmol. Soc. 80:758-822, 1982.

717. Srinivasan, B.D., and Kulkarni, P.S. The role of arachidonic acid metabolites in the mediation of the polymorphonuclear leukocyte response following corneal injury. Invest. Ophthalmol. Vis. Sci. 19: 1087-1093, 1980.

718. Srinivasan, B.D., and Kulkarni, P.S. Polymorphonuclear leukocyte response. Inhibition following corneal epithelial denudation by steroidal and non-steroidal anti-inflammatory agents. Arch. Ophthalmol. 99:1085-1089, 1981.

719. Srinivasan, B.D., Kulkarni, P.S., and Eakins, K.E. Characterization of chemotactic factors in corneal wound healing. In Advances in Prostaglandins and Thromboxane Research. Vol. 7. Samuelsson, B., Ramwell, P.W., and Paoletti, R. (editors). New York, Raven Press, 1980, pp 861-864.

720. Stainer, G.A., Brightbill, F.S., Holm, P., and Laux, D. The development of pseudopterygia in hard contact lens wearers. Contact Intraocular Lens Med. J. 7:1-4, 1981.

721. Stein, M.B., Asbell, P.A., Kamenar, T., Brotman, J., Tapper, M., and Robillard, N. Inhibition of corneal neovascularization by combination heparin-steroid therapy. Invest. Ophthal. Vis. Sci. 28(Suppl) 231, 1987.

722. Stenson, W.F., and Parker, C.W. Metabolism of arachidonic acid in ionophore-stimulated neutrophils. Esterification of a hydroxylated metabolite into phospholipids. J. Clin. Invest. 64:1457-1465, 1979.

723. Stern, D.M., Bank, I., Nawroth, P.P., Cassimeris, J., Kisiel, W., Fenton, J.W., II, Dinarello, C., Chess, L., and Jaff, E.A. Self-regulation of procoagulant events on the endothelial cell surface. J. Exp. Med. 162:1223-1235, 1985.

724. Stock, E.L., Mendelsohn, A.D., Lo, G.G., Ghosh, S., and O'Grady, R.B. Lipid keratopathy in rabbits: an animal model system. Arch. Ophthalmol. 103:726, 1985.

725. Stolpen, A.H., Guinan, E.C., Fiers, W., and Pober, J.S. Recombinant tumor necrosis factor and immune interferon act singly and in combination to reorganize human vascular endothelial cell monolayers. Amer. J. Pathol. 123:16-24, 1986.

726. Strohl, D., Russell, W.A., Haynes, W.L., Smith, C.F., and Klintworth, G.K. Quantitation of corneal neovascularization in the rat at various times after chemical cauterization. Invest. Ophthalmol. Vis. Sci. 28(Suppl.):28, 1987.

727. Strydom, D.J., Fett, J.W., Lobb, R.R., Alderman, E.M., Bethune, J.L., Riordan, J.F., and Vallee, B.L. Amino acid sequence of human tumor derived angiogenin. Biochemistry 24:5486-5494, 1985.

728. Sugiura, S., and Matsuda, H. Electron microscopic observations on the corneal neovascularization. Acta Soc. Ophthalmol. Jap. 73: 1208-1221, 1969 (In Japanese).

729. Sugme, S.P., and Hay, E.D. Response of basal epithelial cell surface and cytoskeleton to solubilised extracellular matrix molecules. J. Cell Biol. 91:45-54, 1981.

730. Sulkin, D.F., Sulkin, N.M., and Nushan, H. Corneal fine structure in experimental scorbutus. Invest. Ophthalmol. 11:633-643, 1972.

731. Suvarnamani, C., Halperin, E.C., Proia, A.D., and Klintworth, G.K. The effect of total lymphoid irradiation on corneal vascularization in the rat following chemical cautery. Radiat. Res. 117:259-272, 1989

732. Svensjö, E., and Grega, G.J. Evidence for endothelial cell-mediated regulation of macromolecular permeability by postcapillary venules. Fed. Proc. 45:89-95, 1986.

733. Swindle, P.F. Events of vascularization and devascularization seen in corneas. Arch. Ophthalmol. 20:974-995, 1938.

734. Sydenstricker, V.P., Hall, W.K., Hock, C.W., and Pund, E.R. Amino acid and protein deficiencies as causes of corneal vascularization: A preliminary report. Science 103:194-196, 1946.

735. Szalay, J. Permeability of rat corneal vessels to fluorescein and carbon. Exp. Eye Res. 18:447-456, 1974.

736. Szalay, J., and Pappas, G.D. Fine structure of rat corneal vessels in advanced stages of wound healing. I. Permeability to intravenously injected thorium dixoide. Microvasc. Res. 2: 319-329, 1970.

737. Szalay, J., and Pappas, G.D. Fine structure of rat corneal vessels in advanced stages of wound healing. Invest. Ophthalmol. 9:354-365, 1970.

738. Szeghy, G. Experimentelle Untersuchungen über die Hornhautvaskularisation auslösende Wirkung der Tränenflüssigkeit (Bindehautsekret) von Kaninchen und Menschen, I. Klin. Monatsbl. Augenheilkd 155:57-61, 1969.

739. Tabatabay, C.A., Baumgartner, B., and Leuenberger, P.M. Cellules de Langerhans et néovascularisation cornéenne dans les brûlures alcalines expérimentales. J. Franc. D'Opthalmol. 10:419-423, 1987.

740. Takigawa, M., Shirai, E., Enomoto, M., Pan, H.-O., Suzuki, F., Shiio, T., and Yugari, Y. A factor in conditioned medium of rabbit costal chondrocytes

inhibits the proliferation of cultured endothelial cells and angiogenesis induced by B16 melanoma: its relation with cartilage-derived anti-tumor factor (CATF). Biochem. Int. 14:357-363, 1987.

741. Takigawa, M., Shirai, E., Enomoto, M., Kinoshita, A., Pan, H.-O., Suzuki, F., and Yugari, Y. Establishment from mouse growth cartilage of clonal cell lines with responsiveness to parathyroid hormone alkaline phosphatase activity, and ability to produce an endothelial cell growth inhibitor. Calcif. Tissue Int. 45:305-313, 1989.

742. Talman, E.L., and Harris, J.E. Ocular changes induced by polysaccharides. III. Paper chromatographic fractionation of a biologically active hyaluronic acid sulfate preparation. Am. J. Ophthalmol. 48:560-572, 1959.

743. Tannock, I.F., and Hayaski, S. The proliferation of capillary endothelial cells. Cancer Res. 32: 77-82, 1972.

744. Tavassoli, M., and Crosby, W.H. Transplantation of marrow to extramedullary sites. Science 161:54-56, 1968.

745. Tavassoli, M., and Crosby, W.H. The fate of fragments of liver implanted in ectopic sites. Anat. Rec. 166:143-152, 1970.

746. Tawara. Concerning the capillaries in the cornea of Megalobatrachus maximus (Cryptobranchus japonicus) Nagasaki Igak. Zas. 11: 350-354, 1933.

747. Taylor, C.M., McLaughlin, B., Weiss, J.B., and Smith, I. Bovine and human pineal glands contain substantial quantities of endothelial cell stimulating angiogenic factor. J. Neural. Transm. 71:79-84, 1988.

748. Taylor, S., and Folkman, J. Protamine is an inhibitor of angiogenesis. Nature 297:307-312 1982.

749. Terranova, V.P., DiFlorio, R., Lyall, R.M., Hic, S., Friesel, R., and Maciag, T. Human endothelial cells are chemotactic to endothelial cell growth factor and heparin. J. Cell Biol. 101:2330-2334, 1985.

750. Thakral, K.K., Goodson, W.H., III, and Hunt, T.K. Stimulation of wound blood vessel growth by macrophages. J. Surg. Res. 26:430-438, 1979.

751. Thoft, R.A., Friend, J., and Murphy, H.S. Ocular surface epithelium and corneal vascularization in rabbits. I. The role of wounding. Invest. Ophthalmol. Vis. Sci. 18:85-92, 1979.

752. Thomas, K.A. Fibroblast growth factors. FASEB J 1:434-440, 1987.

753. Thomas, K.A., Rios-Candelore, M., and Fitzpatrick, S. Purification and characterization of acidic fibroblast growth factor from bovine brain. Proc. Natl. Acad. Sci. USA 81:357-361, 1984.

754. Thomas, K.A., Rios-Candelore, M., Gimenez-Gallego, G., DiSalvo, J., Bennett, C., Rodkey, J., and Fitzpatrick, S. Pure brain-derived acidic fibroblast growth factor is a potent angiogenic vascular endothelial cell mitogen with sequence homology to interleukin 1. Proc. Natl. Acad. Sci. USA 82:6409-6413, 1985.

755. Thompson, J.A., Anderson, K.D., Dipietro, J.M., Zwiebel, J.A., Zametta, M., Anderson, W.F., and Maciag, T. Site directed neovessel formation in vivo. Science 241:1349-1352, 1988.

756. Thompson, J.A., Haudenschild, C.C., Anderson, K.D., DiPietro, J.M., Anderson, W.F., and Maciag, T. Heparin-binding growth factor 1 induces the formation of organoid neovascular structures in vivo. Proc. Natl. Acad. Sci. USA 86:7928-7932, 1989.

757. Thompson, W.D., and Brown, F.I. Quantitation of histamine induced angiogenesis in the chick chorioallantoic membrane: mode of action of histamine is indirect. Int. J. Microcirc. Clin. Exp. 6:343-357, 1987.

758. Thompson, W.D., and Campbell, R. Stimulation of angiogenesis by fibrin degradation products - a mechanism proposed for tissue repair and tumour growth. Int. J. Microcirc. Clin. Exp. 1:329, 1982.

94

759. Thompson, W.D., Campbell, R., and Evans, A.T. Fibrin degradation and angiogenesis: quantitative analysis of the angiogenic response in the chick chorioallantoic membrane. J. Pathol. 145:27-37, 1985.

760. Thornton, S.C., Mueller, S.N., and Levine, E.M. Human endothelial cells: use of heparin in cloning and long-term serial cultivation. Science 222:623-625, 1983.

761. Till, G.O., Johnson, K.J., Kunkel, R., and Ward, P.A. Intravascular activation of complement and acute lung injury. Dependency on neutrophils and toxic oxygen metabolites. J. Clin. Invest. 69:1126-1135, 1982.

762. Todderud, G., and Carpenter, G. Epidermal growth factor: the receptor and its function. Biofactors 2:11-15, 1989.

763. Tommila, P., Summanen, P., and Tervo, T. Cortisone heparin and argon laser in the treatment of corneal neovascularization. Acta Ophthalmol. Suppl. 182:89-92, 1987.

764. Toole, B.P. In Neuronal Recognition S.H. Baronades, (Ed.), Plenum, New York, 1976, pp 275-329.

765. Toole, B.P., Chitra, B., and Gross, J. Hyaluronate and invasiveness of the rabbit V2 carcinoma. Proc. Natl. Acad. Sci. USA 76: 6299-6303, 1979.

766. Totter, J.R., and Day, P.L. Cataract and other ocular changes resulting from tryptophane deficiency. J. Nutr. 24:159-166, 1942.

767. Toynbee, J. Researches, tending to prove the non-vascularity and the peculiar uniform mode of organization and nutrition of certain animal tissues, viz. articular cartilage, and the cartilage of the different classes of fibrocartilage; the cornea, the crystalline lens, and the vitreous humour; and the epidermoid appendage. Philos. Trans. R. Soc. Lond [Biol] 131:159-192, 1841.

768. Tseng, S.C.G., Farazdaghi, M., and Rider, A.A. Conjunctival transdifferentiation induced by systemic vitamin A deficiency in vascularized rabbit corneas. Invest. Ophthalmol. Vis. Sci. 28:1497-1504, 1987.

769. Tseng, S.C.G., Hirst, L.W., Farazdaghi, M., and Green, W.R. Goblet cell density and vascularization during conjunctival transdifferentiation. Invest. Ophthalmol. Vis. Sci. 25:1168-1176, 1984.

770. Tsujimoto, M., Yokota, S., Vilček, J., and Weissmann, G. Tumor necrosis factor provokes superoxide anion generation from neutrophils. Biochem. Biophys. Res. Commun. 137:1094-1100, 1986.

771. Tsukamoto K., and Sugino Y. Tumor angiogenesis activity in clonal cells transformed by bovine adenovirus type 3. Cancer Research 39:1305-1309, 1979.

772. Tsuruoka, N., Sugiyama, M., Tawaragi, Y., Tsujimoto, M., Nishihara, T., Goto, T., and Sato, N. Inhibition of in vitro angiogenesis by lymphotoxin and interferon-γ. Biochem. Biophys. Res. Commun. 155:429-436, 1988.

773. Tuan, D., Smith, S., Folkman, J., and Merler, E. Isolation of the nonhistone proteins of rat Walker carcinoma 256: Their association with tumor angiogenesis. Biochemistry 12:3159-3165, 1973.

774. Ungari, S., Katari, R.S., Alessandri, G., and Gullino, P.M. Cooperation between fibronectin and heparin in the mobilization of capillary endothelium. Invasion Metastasis 5:193-205, 1985.

775. Van den Brenk, H.A.S., Sharpington, C., Orton, C., and Stone, M. Effects of x-irradiation on growth and function of the repair blastema (granulation tissue). II. Measurement of angiogenesis in the Selye pouch in the rat. Int. J. Rad. Biol. 25:277-289, 1974.

776. Van Haeringen, N.J., Van Delft, J.L., Barthen, E.R., de Wolff-Rouendaal, D., and Oosterhuis, J.A. Effect of indomethacin on immunogenic keratitis. Curr. Eye Res. 5:307-311, 1986.

777. Verbey, N.L.J., van Haeringen, N.J., and de Jong, P.T.V.M. Modulation of immunogenic keratitis in rabbits by topical administration of poly-unsaturated fatty acids. Curr. Eye Res. 7:549-556, 1988.

778. Vigny, M., Ollier-Hartmann, M.P., Lavigne, M., Fayein, N., Jeanny, J.C., Laurent, M., and Courtois, Y. Specific binding of basic fibroblast growth factor to basement membrane-like structures and to purified heparan sulfate proteoglycan of the EHS tumor. J. Cell. Physiol. 137:321-328, 1988.

779. Vlodavsky, I., Folkman, J., Sullivan, R., Fridman, R., Ishai-Michaeli, R., Sasse J., and Klagsbrun, M. Endothelial cell-derived basic fibroblast growth factor: Synthesis and deposition into subendothelial extracellular matrix. Proc. Natl. Acad. Sci. USA 84:2292-2296, 1987.

780. Vlodavsky, I., Johnson, L.K., Greenburg, G., and Gospodarowicz, D. Vascular endothelial cells maintained in the absence of fibroblast growth factor undergo structural and functional alterations that are incompatible with their in vivo differentiated properties. J. Cell Biol. 83: 468-486, 1979.

781. Vlodavsky, I., Klagsbrun, M., and Folkman, J. Storage of heparin-binding endothelial cell growth factors in the cornea: a new mechanism for corneal neovascularization. In Ocular Circulation and Neovascularization. BenEzra, D., Ryan, S.J., Glaser, D.M., and Murphy, R.P. (Eds), Martinus Nijhoff/Dr. W. Junk, Dordrecht, 1987.

782. Vu, M.T., Burger, P.C., and Klintworth, G.K. Angiogenic activity in injured rat corneas as assayed on the chick chorioallantoic membrane. Lab. Invest. 53:311-319, 1985.

783. Vu, M.T., Smith, C.F., Burger, P.C., and Klintworth, G.K. Methods in Laboratory Investigation: An evaluation of methods to quantitate the chick chorioallantoic membrane assay in angiogenesis. Lab. Invest. 53:499-508, 1985.

784. Wagner, C.R., Vetto, R.M., and Burger, D.R. The mechanism of antigen presentation by endothelial cells. Immunobiology (Stuttgart). 168:453-469, 1984.

785. Wahl, S.M., Hunt, D.A., Wakefield, L.M., McCartney-Francis, N., Wahl, L.M., Roberts, A.B., and Sporn, M.B. Transforming growth factor type β induces monocyte chemotaxis and growth factor production. Proc. Natl. Acad. Sci. USA 84:5788-5792, 1987.

786. Wahl, S.M., Wong, H.L., and McCartney, F.N. Role of growth factors in inflammation and repair. J. Cell Biochem. 40:193-199, 1989.

787. Waldrep, J.C., and Crosson, C.E. Induction of keratouveitis by capsaicin. Current Eye Res. 7:1173-1182, 1988.

788. Wall, R.T., Harker, L.A., Quadracci, L.J., and Striker, G.E. Factors influencing endothelial cell proliferation in vitro. J. Cell. Physiol. 96:203-213, 1978.

789. Wall, R.T., Harker, L.A., and Striker, G.E. Human endothelial cell migration. Stimulation by a released platelet factor. Lab. Invest. 39:523-529, 1978.

790. Walsh, C.E., Dechatelet, L.R., Thomas, M.J., O'Flaherty, J.T., and Waite, M. Effect of phagocytosis and ionophores on release and metabolism of arachidonic acid from human neutrophils. Lipids 16:120-124, 1981.

791. Wang, H., Berman, M., and Law, M. Latent and active plasminogen activation in corneal ulceration. Invest. Ophthalmol. Vis. Sci. 26:511-524, 1985.

792. Waterbury, L.D., Zagotta, M.T., and Shott, L.D. The effect of dexamethasone on corneal neovascularization induced by epidermal growth factor. Proc. West. Pharmacol. Soc. 24:153-158, 1981.

793.   Watt, S.L., and Auerbach, R.   A mitogenic factor for endothelial cells obtained from mouse secondary mixed leukocyte cultures. J. Immunol. 136: 197-202, 1986.

794.   Weibel, E.R., and Palade, G.E.   New cytoplasmic components in arterial endothelia. J. Cell Biol. 23:101-112, 1964.

795.   Weiss, J.B., Brown, R.A., Kumar, S., and Phillips, P.   An angiogenic factor isolated from tumours: a potent, low-molecular-weight compound. Br. J. Cancer 40: 493-496, 1979.

796.   Weiss, J.B., Hill, C.R., Davis, R.J., McLaughlin, B., Sedowofia, K.A., and Brown, R.A.   Activation of procollagenase by a low molecular weight angiogenesis factor. Biosci. Rep. 3:171-177, 1983.

797.   Weissmann, G. The role of lysosomes in inflammation and disease. Ann. Rev. Med. 18:97-112, 1967.

798.   Welsh, K.I., Burgos, H., and Batchelor, J.R.   The immune response to allogeneic rat platelets: Ag-B antigens in matrix form lacking Ia. Eur. J. Immunol. 7:267, 1977.

799.   Wentzell, B., and Epand, R.M.   Stimulation of the release of prostaglandins from polymorphonuclear leukocytes by the calcium ionophore A23187. FEBS Letters 86:255-258, 1978.

800.   Wessely, K.   Ueber anaphylaktische Erscheinungen an derHornhout. (Experimentelle Erzeugung einer parenchymatösen Keratitis durch artfremdes Serum. Münch Med. Wochenschr. 58:1713-1714, 1911.

801.   West, D.C., Hampson, I.N., Arnold, F., and Kumar, S.   Angiogenesis induced by degradation products of hyaluronic acid. Science 228: 1324-1326, 1985.

802.   West, D.C., and Kumar, S.   Modulation of hyaluronate and its degradation products in vitro. In Tsuchija, M., Makishige, A., Mishima, Y., and Oda, M. (Eds.), Microcirculation - an update.   Elsevier Science Publishers, Amsterdam, Vol. 2:801-802, 1987.

803.   West, D.C., and Kumar, S.   Endothelial cell proliferation and diabetic retinopathy. Lancet 1:715-716, 1988.

804.   West, D.C., and Kumar, S.   Hyaluronan and angiogenesis. Ciba Foundation Symposium, 1989.

805.   Westermark, B., Heldin, C.-H., Ek, B., Johnson, A., Mellstrom, K., Nister, M., and Wasteson, A.   Biochemistry and biology of platelet-derived growth factor. In Growth and Maturation Factors, G. Guroff (Ed.), John Wiley and Sons, New York, 83-87, 1983.

806.   Williams, K.A., Grutzmacher, R.D., Roussel, T.J., and Coster, D.J.   A comparison of the effects of topical cyclosporine and topical steroid on rabbit corneal allograft rejection. Transplant 39:242-244, 1985.

807.   Williamson, J.S.P., DiMarco, S., and Streilein, J.W.   Immunobiology of Langerhans cells on the ocular surface. 1. Langerhans cells within the central cornea interfere with induction of anterior chamber associated immune deviation. Invest. Ophthalmol. Vis. Sci. 28:1527-1532, 1987.

808.   Windt, M.R., and Rosenwasser, L.J. Lymphokine Res. 3:281(abst), 1984.

809.   Wirsching, L.   Eye symptoms in acrodermatitis enteropathica.   Acta Ophthalmologica 40:567-574, 1962.

810.   Wise, G. Ocular rosacea. Am. J. Ophthalmol. 26:591-609, 1943.

811.   Wiseman, D.M., Polverini, P.J., Kamp, D.W., and Leibovich, S.J.   Transforming growth factor-β (TGF-β) is a chemoattractant for monocytes and induces their expression of angiogenic activity. J. Cell Biol. 105:163a, 1987.

812.   Wissler, J.H., and Renner, H.   Inflammation, chemotropism and morphogenesis: novel leukocyte-derived mediators for directional growth of

blood vessels and regulation of tissue neovascularization. Zeit. Physiol. Chem. 362:244, 1981.

813. Wolbach, S.B., and Howe, P.R. Tissue changes following deprivation of fat soluble A vitamin. J. Exp. Med 42:753-777, 1925.

814. Wolf, J.E., and Harrison, R.G. Demonstration and characterization of an epidermal angiogenic factor. J. Invest. Dermatol. 61:130-141, 1973.

815. Wood, S., Lewis, R., Mulholland, J.H., and Knaack, J. Assembly, insertion and use of modified rabbit ear chamber. Bull. Hopkins Hosp. 119:1, 1966.

816. Wybar, K.C., and Campbell, F.W. The influence of cortisone on corneal vascularization in the guinea pig and in the rabbit. Trans. Ophthalmol Soc. UK 72:105-117, 1952.

817. Yamagami, I. Electron microscopic study of the cornea. I. The mechanisms of experimental new vessel formation. Acta Soc. Ophthalmol. Jap. 73: 1222-1242, 1969.

818. Yamagami, I. Electron microscopic study of the cornea I. The mechanism of experimental new vessel formation. Jpn. J. Ophthalmol. 14:41, 1970.

819. Yamamoto, S. Enzymes synthesizing and metabolizing prostanoids. In Biochemistry of Arachidonic Acid Metabolism, Lands, W.E.M. (Ed.), Martinus Nijhoff, Boston, 1985.

820. Yamane, A., Tokura, T., Nishikawa, M., Ito, S., and Miki, H. Morphological study of experimental corneal neovascularization after anterior uveal ischemia [Japanese]. Acta Soc. Ophthalmol. Jap. 93:315-321, 1989.

821. Yasunaga, C., Nakashima, Y., and Sueishi, K. A role of fibrinolytic activity in angiogenesis. Quantitative assay using in vitro method. Lab. Invest. 61:698-704, 1989.

822. Zauberman, H., Michaelson, I.C., Bergman, F., and Maurice, D.M. Stimulation of neovascularization of the cornea by biogenic amines. Exp. Eye Res. 8:77-83, 1969.

823. Zauberman, H., and Refojo, M.F. Keratoplasty with glued-on lenses for alkali burns: an experimental study. Arch. Ophthalmol. 89: 46-48, 1973.

824. Zetter, B.R. The endothelial cells of large and small blood vessels. Diabetes 30(Suppl. 2): 24-28, 1981.

825. Zetter, B.R. Angiogenesis: state of the art. Chest 93 (Suppl.) 159, 1988.

826. Zetter, B.R., and Antoniades, H.N. Stimulation of human vascular endothelial cell growth by a platelet-derived growth factor and thrombin. J. Supramol. Struct. 11:361-370, 1979.

827. Ziche, M., Alessandri, G., and Gullino, P.M. Gangliosides promote the angiogenic response. Lab. Invest. 61:629-634, 1989.

828. Ziche, M., Banchelli, G., Caderni, G., Raimondi, L., Dolara, P., and Buffoni, F. Copper-dependent amine oxidases in angiogenesis induced by prostaglandin E_1 (PGE_1). Microvasc. Res. 34:133-136, 1987.

829. Ziche, M., Jones, J., and Gullino, P.M. Role of prostaglandin E_1 and copper in angiogenesis. J. Natl. Cancer Instit. 69:475-482, 1982.

830. York, K., Gomer, C.J., Murphee, A.L., and Schanzlin, D.J. Hematoporphyrin derivative photoradiation therapy (HPD-PRT) for corneal neovascularization. Invest. Ophthalmol. Vis. Sci. 26(Suppl):180, 1985.

831. Yoshizuka, M., Annoura, S., Yagyu, A., and Baba, K. Causative factors of the cornea neovascularization. Kurume Med. J. 26: 15-20, 1979.

832. Zurier, R.B., and Weissman, G. Anti-immunologic and anti-inflammatory effects of steroid therapy. Med. Clin. N. Amer. 57:1295-1307, 1973.

INDEX

A

acetylcholine 37, 42, 52

acidic fibroblast growth factor (aFGF) 36, 46

adenosine 6

adenosine 3'-5' monophosphate (cAMP) 52

adenosine diphosphate (ADP) 6, 15

adenosine triphosphate (ATP) 34

adenylatecyclase 52

adrenocortical cells 50

albumin 21

alkali-burned cornea 4, 16, 17, 27, 30

alloxan 10, 27, 30

α fibroblast growth factor 1 40

α interferon 41

α transforming growth factor 45

α-retina-derived growth factor 46

Alzet minipump 13

angiogenesis 34

angiogenesis, suppression 16

angiogenic factors 25

angiogenin 45

angioimmunoblastic (immunoblastic) lymphadenopathy 33

angiotropin 34, 36, 41

ankylosing spondylitis 45

anti-angiogenic factors 23, 24

anti-mouse-platelet serum 21

antigen 4

antigens 27, 30

antihistamine 6

antimitotics 19

antineutrophil serum 31

antiprotease 20

aorta 21

arachidonic acid 17, 32, 38

argon laser photocoagulation 10

armadillo 2

arteries 8

arterioles 8, 9

arthritis 45
ascorbic acid deficiency 4
autacoids 6
autocrine 43
autoradiography 5, 8

B

basal lamina 9, 37
basement membrane collagen 44
basic fibroblast growth factor (bFGF) 37, 41, 43, 46, 47
benzalkonium chloride 24
benzylamine oxidase 39
beta irradiation 19
β-phorbol 12,13-dibutyrate (PDBu) 52
β-phorbol 12-myristate 13-acetate (PMA) 39
β2 receptor agonists 6
bioassay system for angiogenic factors 13
biogenic amines 36
blood vessels, ultrastructure of growing 5
bovine brain derived growth factor (a HPGF-1) 46
bovine retinal extract 7
Boyden chamber 38
BP 961 39
bradykinin 6, 37, 43
brain 45
brilliant benzo blue 6A 6
bullous keratopathy 24
burns, hyperthermal 30
BW 755C 18
BW A540C 18

C

c-erb B 47
c-sis 48
C755 mammary tumors 3
calcium, intracellular 52
cAMP 52
capillaries 8
capillary regression 9
carbon 6
cartilage 14, 20, 24

cell cycle 7, 8
chemical 4
chemical cautery 5
Chicago blue 6
chick chorioallantoic membrane (CAM) assay 13, 14, 17, 20, 26, 34-36, 39, 41, 46, 47, 49,
    52, 53
chondrocytes 50
chondroitin sulfate lyases 20
chondrosarcoma-derived endothelial cell growth factor 46
choroid 45
class II histocompatibility antigens 41
colchicine 27, 30
collagen 43, 44
collagen, type IV 9
collagenase 7, 37, 39, 47, 53
columbinic acid 21
computerized planimetry 12
concanavalin A 33, 34
congenital endothelial corneal dystrophy 24
conjunctival transdifferentiation 1
contact lenses 4
copper 16, 35, 39, 41, 45
cornea to assay for angiogenic activity 13
corneal avascularity 1
corneal burns 1
corneal clouding 1
corneal edema 24
corneal endothelial cells 50
corneal endothelium, disorders 24
corneal epithelium 37, 40
corneal fibroblasts 23, 33
corneal grafts 27
corneal vascularization, quantitation 11
corneal vascularization, theories 23
corpus luteum 45
corticosteroids 16, 31, 32, 36, 38
cortisone 6, 16, 36
cryotherapy 10
cyclooxygenase 17, 18, 21, 32, 38
cyclooxygenase inhibitors 27, 38

Langerhans' cells 9
leishmaniasis 4
leukocytes 9, 26-28, 30, 31, 41, 44, 50
leukopenia 18
leukotriene 17, 38, 39
leukotriene B4 6, 39
leukotriene C4 6, 39
leukotriene D4 6, 39
leukotriene E4 6
light microscopy 5
lipocortin 17
lipooxygenase 17, 32, 38
lipooxygenase, inhibitors 18
Listeria monocytogenes 34
lizard 2
lymph nodes 2, 10, 19
lymphatic vessels 10
lymphoblast 20
lymphocytes 9, 17, 20, 23, 33, 38, 43, 45
lymphocytoma 33
lymphokines 33
lymphoplasia, cutaneous 33
lysosomal membranes 17

M
macrophage derived growth factor 34, 36, 49
macrophages 17, 26, 33, 34, 38, 40, 41, 45
manatee 2
mast cell secretagogue 35
mast cells 9, 19, 35, 37
medroxyprogesterone 17
Megalobatrachus maximus 2
melanocytes 9, 50
melanoma 14
mepyramine maleate 37
mesodermal growth factor 36
methacrylate casts 9
methylcellulose 13
methylmethacrylate casts 5, 7, 8
methylprednisolone 16
Metiamide Clemastine fumarate 37

mice 19
mice, WWv mutant 35
microvasculature 42
mineralocorticoid 17
monocyte/macrophage derived growth factor (MDGF) 34, 40
monocytes 9, 34, 40, 43, 44
monocytopenia 17
mouse 2, 9, 13, 40
mouse adipocyte conditioned medium 7
mucopolysaccharides 23

N
neoplasms 26, 45
neoplastic cells 44
neovascularization, iris 1
neutrophil leukocytes 31, 32
nicotinamide 46
non-steroidal anti-inflammatory drugs 16, 18, 31, 32, 38
nutritional deficiency states 4

O
onchocerciasis 4
oncogenes 46, 52
opacification 1
osteoarthritis 20
osteoblasts 50
ovary 45
ovary, luteinization 1
Ovis canadensi Shaw 2
oxygen 21, 26, 27

P

pancreatic ribonuclease A 48
pannus 1, 2
pannus degenerativus 2
pannus leprosus 2
pannus phlyctenulosus 2
pannus trachomatosus 2
PD-ECGF 34, 48
pericytes 8
permeability of blood vessels 17

procollagenase 43
prostacyclin 38
prostaglandin 18, 32
prostaglandin E_1 6, 17, 38
prostaglandin E_2 6
prostaglandin $F_{2\alpha}$ 6
prostaglandin synthetase 18
prostaglandins 17, 33, 38, 42, 45
prostnomelysin 43
protamine 19, 35
protamine sulfate 35
proteases 32, 34, 35, 37, 43, 46
protein kinase C (PKC) 52, 53
protein kinases 52

Q
quercetin 18

R
rabbit 5, 7, 9, 10, 13, 17, 19, 32, 35, 40, 42
rabbit ear chambers 8, 13
radiation 14
radiofrequency burns 30, 37
rat 2, 5-7, 9, 13, 18, 37, 52
rat kidney capsule model of angiogenesis 13, 47
rats, zinc deficient 4
receptors 43
receptors, muscarinic cholinergic 52
reptile spectacle 1, 2
retina 44
    neovascularization 26
    pigment epithelium 44
retinoblastoma 14
retinoic acid 4
retinopathy of prematurity 1, 26
retinopathy, proliferative diabetic 26
REV 5901 18
rheumatoid arthritis 20
riboflavin 26, 28
riboflavin-deficient 30, 31

Rocky mountain bighorned sheep 2
rose bengal 10

S
S180 sarcomas 3
salamanders 2
saline 4
salivary glands 45
scanning electron microscopy 5, 8, 9
scurvy 21
Selye pouch 19
serine proteases 36, 37
serotonin 36, 37
serum 45
sialic acid 5, 50
silica particles 19
silicone 26
silver/potassium nitrate 4, 5, 9, 16, 18, 19, 21, 27, 30
skin 44
smooth muscle 40, 49, 50, 53
sodium hydroxide 25, 27
sodium salicylate 18, 32
starvation 4
steroids 16
sterology 12
Stevens-Johnson's syndrome 4
stromal edema 6, 24
subcutaneous air "pouch" of Selye 13
substance P, 6
^{35}S-sulfate 23
superoxide 33

T
T lymphocytes 33
tetradecanoyl phorbol-13-acetate 53
TGF-α 34
TGF-β 34, 48, 49
TGF-β1 48
TGF-β2 48, 49
TGF-β3 48
TGF-β4 48

thermal burns 8, 17, 27, 28, 32
Thiotepa 19
thorium dioxide 6
3T3 fibroblasts 49, 53
3T3-adipocytes 45
thrombin 37
thromboxanes 38
ticabesone propionate 16
tissue factor 42
toxins 4
trachoma 4
transforming growth factor beta (TGF-β) 45, 48
transplanted tissues 1
traumatic 4
triamcinolone 16
triethylene thiophosphoramide 19
tumor 1, 9
tumor cells 8, 14
tumor necrosis factor (TNF) 17, 41, 46
tumor necrosis factor α (TNFα) 34, 36, 41
tumor necrosis factor β 34
12-HETE 39
12-O-tetradecanoyl phorbol-13-acetate) (TPA) 7
2-pyridylethylamine dihydrochloride 37
tyrosine 52
tyrosine protein kinases 52

U
ulcers 1
urokinase 37, 43

V
V2 carcinoma 19
vascular endothelial cell cultures 13
vascular endothelial cells 7
vascular maturation 9
vascular patterns 2
vascular permeability factor 45, 49
vascularization, interstitial 2
vasodilatation 9
vasopressin 6

111

venules 8, 9
venules, postcapillary 6, 8
vitamin A 30
vitamin A deficiency 4, 5, 28
vitreous 20, 45
von Willebrand factor 5, 42
VPF 50

W
Walker carcinoma 16
Weibel-Palade bodies 5
wound healing 1
wound hormone 41
wounds, vascularizing 26

X
x-irradiation 8, 31
xanthines 6

Tables

TABLE 1. Causes of Experimentally Induced Corneal Vascularization
  *Known to be associated with inflammation

Model	References
CHEMICALS	
Chemicals (Intracorneal)	
Heparin (in Elvax pellet) (100 µg/ml)	521,522
Heparin complexed to copper	8,9,617
Selenomethionine (in Elvax pellet)*	520
Phorbol esters*	551
Selenomethionine* (in Elvax pellet)	520
Adenosine diphosphosphate (in Elvax pellet) (10^{-5}M)*	520,522
Gly-His-Lys complexed to copper	9,617
Ceruloplasmin	8,9,617,829
Silica*	748
Dispirin®	515
Ethanol (repeated)	191
Acids	
Acetic acid	374
Hydrochloric acid	143,374
Sulfated derivative of hyaluronic acid	742
Alkalis	
Sodium hydroxide*	143,209,459
Serum (non-autologous)	18,92,195
Serum (autologous)	195
Nicotinamide (in Elvax pellet)	436
Metabolites of Glycolysis	
Lactic acid	374
Pyruvic acid	374
Enzymes	
Collagenase*	468
Hyaluronidase	532
Plasminogen activator*	63,297
Chemicals (Topical)	
Alloxan*	4,18,20,146,263,319,343,446,447,504,580, 706,728
Sodium hydroxide*	101,263,266,393,452,459,565,590,624,739
Silver nitrate*	149,150,251,263,264,494,612,666,667, 692,736,737
Mustard gas	496
Iodine	707
Colchicine*	343,439,525
Chemicals (intracameral)	
Rabbit Eye	
Alloxan	19,155
Guinea Pig Eye	
Alloxan	343
Chemicals (Systemic)	
Colchicine	439
Chloroquine	254

115

TABLE 1. Causes of Experimentally Induced Corneal Vascularization (continued)

Model	References
Inflammatory Mediators	
Biogenic amines	
Acetylcholine	823
Histamine	823
Serotonin	823
Bradykinin	823
Lymphokines	
Interleukin 1*	59,178,340
Interleukin 2*	201,202,205,340
tumor necrosis factor α*	258,456
transforming growth factor β*	224
Prostaglandins	
PGE$_1$*	55,128,468,617,713,829
PGE$_2$	55,617,827
PGE$_1$ and gangliosides	827
PGF$_{2\alpha}$	55
Leukocyte Attractants	
Formyl methionyl leucyl phenylalanine (in Elvax)*	522
Copper sulfate*	513,515,522
Copper sulfate (in Elvax pellet)*	593
Miscellaneous	
Fibrin* (commercial and autologous)	424
Water	143
Nitrogen	18
Physiologic saline	18,195
Phosphate buffered saline* (some cases)	195
Growth Factors	
Angiogenin (in methylcellulose)	223
Angiotropin	356,812
Epidermal growth factor	55,306,792
Acidic fibroblast growth factor	474
Acidic fibroblast growth factor and heparin	344
Basic fibroblast growth factor	258,306,314,344,686,827
Basic fibroblast growth factor and gangliosides	827
Bovine endothelium stimulating factors	522
Vascular permeability factor (high doses*)	154
MICROORGANISMS AND THEIR PRODUCTS	
Herpes simplex*	143,341
Intracorneal herpes simplex (after sensitization with	
herpes virus type 2)	450
Mycobacterium tuberculosis	465,478
Aspergillus fumigatus*	249
Staphylococcus	327
Mycobacterium tuberculosis (heat-killed)	465,629
Bacterial nucleoprotein	390
Endotoxin (lipopolysaccharide)*	247,363,557
PHYSICAL INJURIES	
Thermal burn*	116,143,152,229,354,469,597,636,664,
	737,831
Corneal abrasion	359
Hypothermal injury	507

116

TABLE 1. Causes of Experimentally Induced Corneal Vascularization (continued)

Model	References
Radiation	
Radiofrequency burns*	419,579
Carbon dioxide laser irradiation	225
Creation of cleavage plane in cornea with probe	18
Deepithelialization with trephination	572
Removal of corneal endothelium and Descemet's	
membrane with blunt curette	18
DEFICIENCY STATES	
Amino acids	
Lysine	5,113,355,734,766
Methionine	60,355,734
Tryptophan	5,6,113,355,766
Histidine	509
Heavy metals	
Zinc	248,462
Protein	330
Total starvation	734
Vitamins	
Vitamin A*	90,813
Riboflavin*	65,90,263,810
Ascorbic acid	730
INTOXICATIONS	
Thallium	113
Tyrosine*	112,287,369,419
IMMUNOLOGIC REACTIONS	
Intracorneal antigen in sensitized animal*	263,279,389-391,800
Topical antigen to traumatized cornea of sensitized	
animal*	640
Corneal allograft rejection*	13,345,377,407
Intravitreal injection of antigen*	359,377
Human serum albumin*	777
Intracorneal bovine gamma globulin*	533
Intracorneal bovine serum*	586
INTRACORNEAL INSTILLATION OF CELLS, TISSUES OR	
THEIR COMPONENTS	
1. **Rat Cornea (Leukopenic Irradiated)**	
Polymorphonuclear leukocytes	265
Extracts of polymorphonuclear leukocytes	265
2. **Rat Cornea (Non-Irradiated)**	
Walker 256 carcinoma angiogenic factor	222
Walker 256 carcinoma extract	515
Endothelial stimulating factor (partially purified	
from bovine parotid gland	515
Heparin derived growth factor (HDGF) (rat	
chondrosarcoma)	686
Macrophages (tumor associated)	609
Macrophage conditioned culture medium	609

Model	References
Concanavalin A activated human macrophages	426
Endotoxin activated human macrophages	426
Macrophages from human synovial tissue with rheumatoid arthritis	427
Cytokines	
Tumor necrosis factor α*	258,456
3. **Rabbit Cornea (Non-Irradiated)**	
Cartilage extracts (bovine calf)	61
Choroid (autologous)*	215
Conjunctiva (autologous)	215
Corneal extracts (allogeneic)	229
Corneal extracts (normal and vascularizing rabbit)	18
Corpus luteum (autologous)	318,399
Corpus luteum (rabbit)	314
Fat (subcutaneous autologous)	700
Gamma globulin (bovine)*	533
Heart muscle (bovine) (2/3)	399
Iris (autologous)*	215
Kidney (bovine) (1/3)	399
Kidney (11 day embryonic mouse)	632
Lacrimal fluid	738
Lymph node (rabbit)	748
Lymphocytes (allogeneic)	203
Macrophages	
Culture medium with 15 mM lactate	385
Culture medium with 25 mM lactate	385
Macrophages (autologous)	550
Macrophages (resting wound)	139
Macrophages (stimulated peritoneal)	139
Macrophages (supernatants of cultured activated porcine	812
Muscles (autologous extraocular)	215
Neoplasms	
Brown-Pearce epithelioma	283,285
Chondrosarcoma-derived factor	686
Ependymoblastoma (mouse)	94,95
Glioblastoma (human)	94
"L cell" tumor (murine)	399
Melanoma (viable dog)*	654
Melanoma (viable hamster-Greene)*	654
Melanoma (viable human)*	654
MTW9A carcinomas (interstitial fluid and ethanol extracts)	829
Plasma membrane components of human tumors	7

TABLE 1. Causes of Experimentally Induced Corneal Vascularization (continued)

Model	References
Rabbit V2 carcinoma	25,93,95,218,226,283,285,286,442,611, 748
Rabbit V2 carcinoma (irradiated)	25
Rat 7,12-dimethyl-benz(α) anthracene tumors	617
Retinoblastoma (devitalized Y79 cell culture)*	654
Retinoblastoma (viable human sporadic)*	654
Retinoblastoma (viable Y79 cell culture)*	654
Walker carcinoma (interstitial fluid and ethanol extracts)	829
Nonhistone proteins from malignant cell nuclei	773
Omental extract (cat)*	299,613
Omentum (autologous)	700
Optic nerve (autologous)	215
Ovarian follicle fluid (human)	260
Parotid gland extracts (bovine)	513
Platelets (thrombin-activated autologous)*	424
Retina (extract)	131
Retina (vascular autologous)*	215
Serum (horse)	776
T lymphocytes (supernatants of activated syngeneic)	482
Vitreous (bovine and human)	218
4. Rabbit Cornea (irradiated)	
Corneal epithelium (perfusion of cultured)	195
Corneal epithelium (allogeneic fresh homogenate)	195
Melanoma (hamster Greene)*	654
Melanoma (devitalized, dog)*	654
Melanoma (devitalized, hamster-Greene)*	654
Retinoblastoma (Y-79 cell culture)*	654
5. Mouse Cornea	
Lymph node fragments (semi-allogeneic)	563
Lymphocytes (allogeneic)	563
Lymphocytes (allogeneic Con A stimulated)	204,205
Macrophage cell line (P388D$_1$)	396
Macrophages (activated syngeneic)	396
Macrophages (resident syngeneic)	396
Splenocytes (syngeneic)	396
Tumors	
C755 mammary tumors	563
P815 mastocytoma cells	807
S180 sarcoma	563
6. Guinea pig cornea	
Macrophages (autologous, homologous and heterologous activated peritoneal)	607

TABLE 1. Causes of Experimentally Induced Corneal Vascularization (continued)

Model	References
Miscellaneous	
Corneal epithelial denudation	572,716
Implants in hamster cheek pouch*	416
Contact lens - long-term use	179,180,183
Contact lens after radial keratotomy	401,402
Encirclement of the eyeball by a rubber tube*	4,227,733
Intracorneal sutures	156,333,529,538,592,644,724
Needle tracts penetrating into anterior chamber	18
Full-thickness keratoplasties (with donor material from normal or vascularized corneas)	18
Corneal homografts transplanted into vascularized skin	79
Anterior segment ischemia*	539,615,820
Injuries to cornea with conjunctival epithelium	751
Boviserin-Ferritin	358
Intracorneal perfusion of TC 199 (with or without added serum)*	195
Intracorneal perfusion of Eagle's essential medium (with and without added serum)*	195
Retrobulbar capsaicin*	787
Subcutaneous xylazine and ketamine*	433
Spontaneous Animal Models	
DBA/2 mice	206

TABLE 2. Chronology of corneal response to silver/potassium nitrate cautery in rats

Time following Injury	Event
1 hour	Dilation of pericorneal blood vessels
4 hours	Diapedesis of leukocytes, increased vascular permeability
6 hours	Dilation of pericorneal lymphatics, contact of inflammatory cells with cautery site, corneal clouding
18 hours	Enlargement of vascular endothelial cells and pericytes, first increased ^{3}H-thymidine labeling of corneal epithelial cells
21 hours	First ^{3}H-thymidine labeling of vascular cells and corneal endothelial cells
27 hours	First buds from pericorneal venules and capillaries
36 hours	Mitoses in vascular endothelial cells
42 hours	Many new vessels emerging from pericorneal vessels
69 hours	Corneal re-epithelialization complete
72-75 hours	Return to control levels of labeling indices of corneal epithelium and endothelium
7 days	Regression of many new vessels, return to control level of vascular endothelial cell labeling index, clearing of cornea
9 days	Venules and arterioles first evident within cornea
30 days	Clear cornea, inflammation absent, few loops of large afferent and efferent vessels between cautery site and pericorneal vascular plexus

Table 3. Substances Instilled into Cornea and Reported Not to Cause Neovascularization

Cornea	Intracorneal Substance	Reference
	Endotoxin	139
	Granulocytes (resting and stimulated)	139
	Hank's buffered saline	139
	Lymphocytes (resting and stimulated)	139
	Macrophages (resting peritoneal)	139
	Saline	61,419
Rabbit (Normal)	Adenosine diphosphate (in Elvax pellet) (5×10^{-5}M)	522
	Apoceruloplasmin	8,9
	Cartilage (Papain digests) (bovine calf)	61
	Fluid from human ovarian follicles (heated)	260
	Gly-His-Lys	9,617
	Growth Factors	
	Fibroblast Growth Factor (1 µg)+	55
	Epidermal Growth Factor (1 µg)+	55
	Nerve Growth Factor	55
	Heparin	9,617
	Heparin (in Elvax pellet) (25 µg/ml)	522
	Hyaluronidase	537
	Hydron and human chorionic gonadotropin	260
	Hydron and saline	260
	Lymph node tissue (autogenic)	399
	Macrophages (resting peritoneal)	139
	Placenta	
	Elvex-40	55
	Prostaglandins	
	PGE2	9,829
	PGD2	55
	PGA1	55
	Synthetic Peptides	
	formyl-methyonyl-phenylalanine	55
	formyl-methyonyl-leucine-phenylalanine	55
	methyonyl-leucine	55
	methyonyl-phenylalanine	55

Table 3. Substances Instilled into Cornea and Reported Not to Cause Neovascularization (continued)

Cornea	Intracorneal Substance	Reference
	Hydron Pellets	
	Macrophages	
	Culture medium	385
	Culture medium with 1.5 mM lactate	385
	Culture medium with pH 7.4	385
	Culture medium with pH 7.0	385
	Culture medium with pH 6.7	385
	Culture medium with pH 6.2	385
	Culture medium with 1.5 mM pyruvate	385
	Culture medium with 15 mM pyruvate	385
	Culture medium with 25 mM pyruvate	385
	Dulbecco's modified Eagle's medium	385
	containing 250 mM lactate	385
	with 15 mM lactate	385
	with 15 mM lactate	385
	Ovarian follicles (rabbit autologous and homologous)	318
	Platelets (unactivated, autologous)	424
	Platelets (thrombin-activated, washed, autologous)	424
	Platelets (thrombin-activated, washed with added thrombin, autologous)	424
	Polymorphonuclear leukocytes (rabbit peritoneal)	139
	Polymorphonuclear leukocytes (autologous, from peritoneal exudate)	550
	Methionine (in Elvax pellet)	520
	Retina (peripheral rabbit)	215
	Retina (boiled vascular rabbit)	215
	Cornea (autologous rabbit)	215
	Brain tissue	
	cortex, normal (human)	94
	cortex, atrophic (human)	94
	cortex, normal (mouse)	94
	cortex, fetal (rabbit)	94
	glioblastoma, boiled (human)	94
	sclera (autologous rabbit)	215
	muscle (autologous skeletal)	700
	Liver (autologous)	700
	Kidney mesenchyme (uninduced 11 day mouse embryos)	321,632
	Lymphocytes (non-activated)	265
Rat (Irradiated)	Peripheral blood monocytes (human)	426
Rat (non-irradiated)	Albumin	154

123

Table 3. Substances Instilled into Cornea and Reported Not to Cause Neovascularization (continued)

Cornea	Intracorneal Substance	Reference
Guinea Pig	Polymorphonuclear leukocytes	607
	Lymphocytes (activated)	607
	Paraffin oil	607
	Thioglycollate	607
	Peptone	607
	Medium (M199)	607
	Culture medium (RPMI 1640±5% fetal calf serum)	607
	Homologous peritoneal macrophages (non-activated)	607
	Homologous peritoneal macrophages (activated)	607
	Isologous peritoneal macrophages (activated)	607
	Heterologous peritoneal macrophages (non-activated)	607
	Heterologous peritoneal macrophages (incubated in vitro) with latex	607
	Heterologous peritoneal macrophages (incubated in vitro) without latex	607
	Hydron and macrophage conditioned media	607
	Hydron and control media	607
	Hydron carrier	607
	Apoceruloplasmin	617
	Heparin	617
	Gly-His-Lys	617
	Elvax-40 pellet	617
	Saline (0.9%) with or without potassium chloride (rarely)	195
Inbred Mice	Mammary gland	559
	Lymph nodes (syngeneic)	559

- not stated
+ Higher doses induce angiogenesis

124

Table 4: Variable/Disputed Neovascularization Following Inoculations into Rabbit Cornea

Cell Type	Source	Incidence	Reference
Neutrophils	Sterile peritoneal exudates	1/15	286
Lymphocytes	(Spleen and lymph nodes) - normal and PPD-sensitized donors		
Fetal rabbit skin fibroblast	Culture medium (2% O_2)	1/8	385
Rabbit brain capillary endothelial cells	Culture medium (2% O_2)	1/10	385
Papain		no/slight vascularization	61
Saline (0.9%) with or without potassium chloride			195
Saline (0.9%)		3/21	195
Prostaglandin E2		1/10	617
Rat mammary gland protein		1/10	617
Muscle (rabbit)		1/4	25

125

Figures

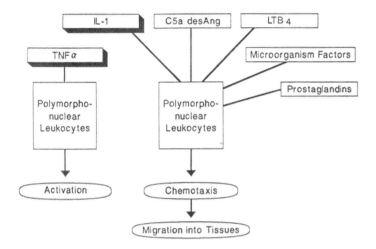

Fig. 1 Some effects on polymorphonuclear leukocytes of cytokines and various other factors.

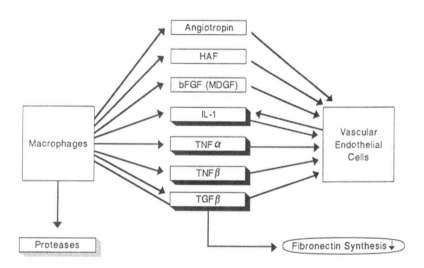

Fig. 2 Some effects of macrophages on vascular endothelial cells.

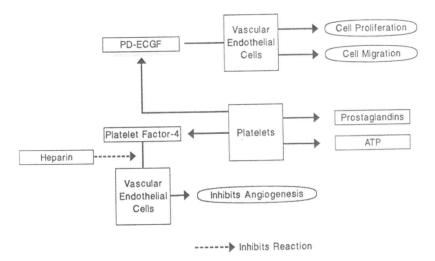

Fig. 3   Some interactions between platelets on vascular endothelium (see also Fig. 4).

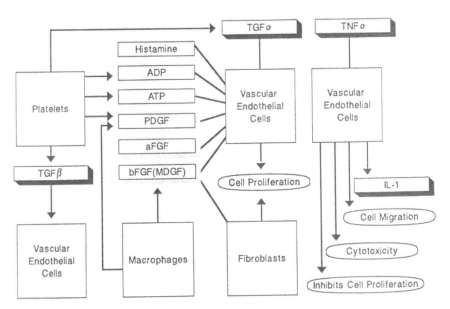

Fig. 4   Effect of certain cytokines, growth factors and histamine on vascular endothelium.

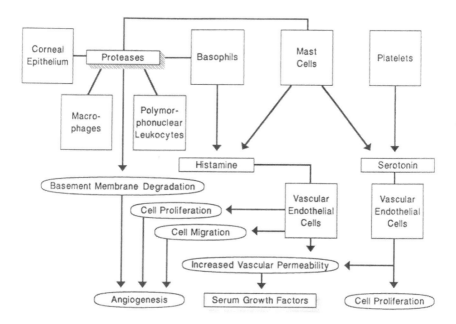

Fig. 5   Relationship between certain cells and the vascular endothelium (see also Figs. 2 and 5).

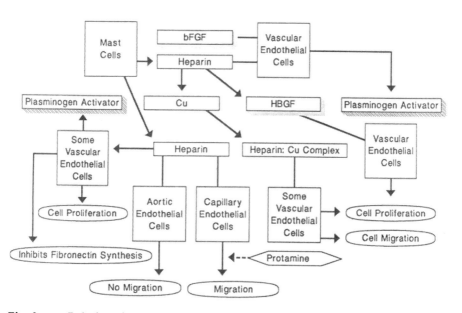

Fig. 6   Relationship between mast cells, heparin and vascular endothelium (see also Fig. 5).

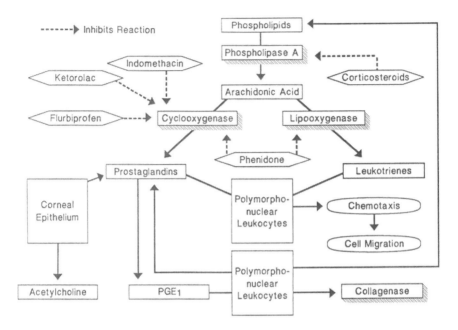

Fig. 7   Effect of eicosanoids and anti-inflammatory drugs on inflammation.

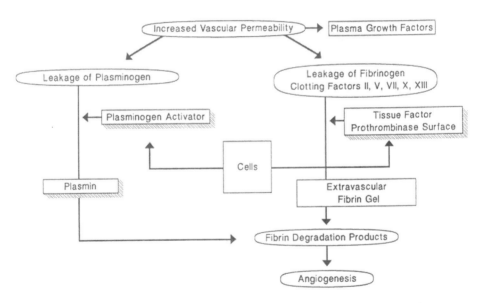

Fig. 8   Influence of increased vascular permeability on angiogenesis.

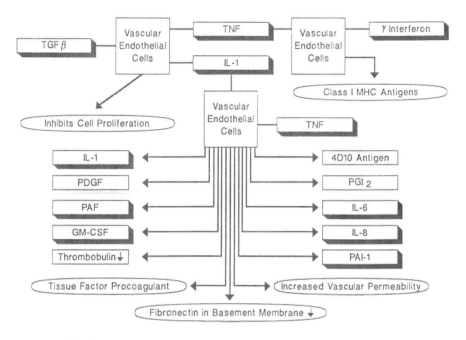

Fig. 9   **Effect of certain cytokines on vascular endothelium.**

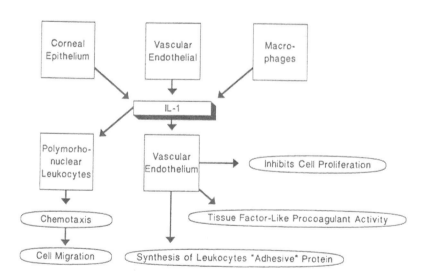

Fig. 10   **Some interactions with Interleukin-1.**

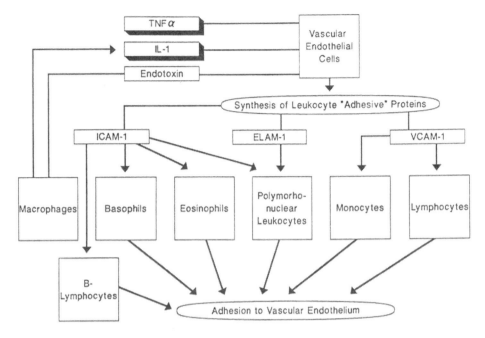

Fig. 11   Mechanism whereby leukocytes adhere to the vascular endothelium.

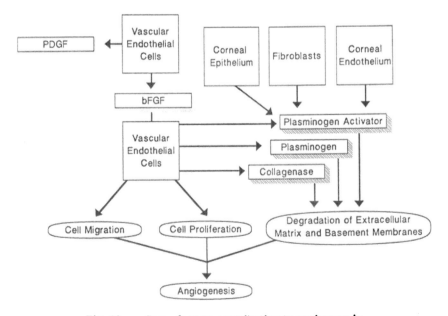

Fig. 12   Some factors contributing to angiogenesis.

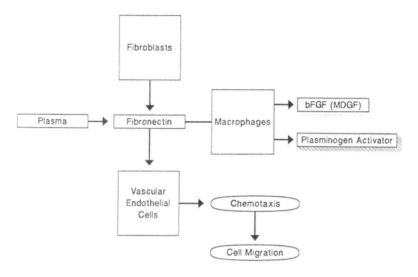

Fig. 13. Fibronectin interactions relevant to angiogenesis.

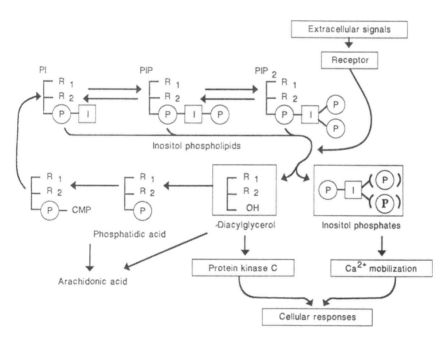

Fig. 14 Depiction of some of the presumed interactions between putative angiogenic factors and the vascular endothelium [Modified from Nishizuka (574)].

CPSIA information can be obtained at www.ICGtesting.com
Printed in the USA
LVOW072355021012

301267LV00002B/11/P